Automobile Wheel Alignment and Wheel Balancing

Automobile Wheel Alignment and Wheel Balancing

R. MANANATHAN

Warrendale, Pennsylvania, USA

400 Commonwealth Drive
Warrendale, PA 15096-0001 USA
E-mail: CustomerService@sae.org
Phone: 877-606-7323 (inside USA and Canada)
724-776-4970 (outside USA)
FAX: 724-776-0790

Copyright © 2022 SAE International. All rights reserved.

No part of this publication may be reproduced, stored in a retrieval system, or transmitted, in any form or by any means, electronic, mechanical, photocopying, recording, or otherwise, without the prior written permission of SAE International. For permission and licensing requests, contact SAE Permissions, 400 Commonwealth Drive, Warrendale, PA 15096-0001 USA; e-mail: copyright@sae.org; phone: 724-772-4028.

Library of Congress Catalog Number 2021948807
http://dx.doi.org/10.4271/9781468603620

Information contained in this work has been obtained by SAE International from sources believed to be reliable. However, neither SAE International nor its authors guarantee the accuracy or completeness of any information published herein and neither SAE International nor its authors shall be responsible for any errors, omissions, or damages arising out of use of this information. This work is published with the understanding that SAE International and its authors are supplying information but are not attempting to render engineering or other professional services. If such services are required, the assistance of an appropriate professional should be sought.

ISBN-Print   978-1-4686-0361-3
ISBN-PDF   978-1-4686-0362-0
ISBN-ePub 978-1-4686-0363-7

To purchase bulk quantities, please contact: SAE Customer Service

E-mail: CustomerService@sae.org
Phone: 877-606-7323 (inside USA and Canada)
724-776-4970 (outside USA)
Fax: 724-776-0790

Visit the SAE International Bookstore at books.sae.org

Chief Growth Officer
Frank Menchaca

Publisher
Sherry Dickinson Nigam

Director of Content Management
Kelli Zilko

Production and Manufacturing Associate
Erin Mendicino

*Dedicated to my loving
wife*
Arul Oli
*(Gracious Light)
who stood by me
during all my research
works
bearing a torch*

Contents

Preface	xi
Foreword by Gurmeet Singh	xiii
Foreword by R. Sivanesan	xv
Foreword by S. Punnaivanam	xvii

CHAPTER 1
Wheel Alignment and Wheel Balancing — 1

Overview	1
Vehicles and Wheels	1
Causes of Tire Wear	2
Wheel Alignment and Wheel Balancing	2

CHAPTER 2
Wheel Balancing — 5

What Is Wheel Balancing?	5
Reasons for Unbalance in Wheels	7
Single Plane Balancing	10
Two Plane Balancing	12
Wheel Balancing Machine—Description	13
Wheel Balancing Machine—Installation	14
Wheel Balancing—Procedure	14
Wheel Balancing—HCVs	17
Wheel Balancing—Two Wheelers	18
Wheel Balancing—Important Points	19

CHAPTER 3

Wheel Alignment of Light Vehicles — 21

Light Vehicles—Definition — 21
Wheel Alignment Angles and Parameters — 21
Parameters affecting Wheel Alignment — 34
Wheel Alignment angles measuring technology and Its advancement — 42
Charge Coupled Device (CCD) Technology — 43
3D Technology — 46
3D Wheel Alignment Computer and Its Accessories — 50
3D Wheel Aligner—Installation — 56
3D Wheel Aligner—Calibration — 59
3D Wheel Alignment Procedure — 62

CHAPTER 4

Wheel Alignment Correction Methods — 69

Wheel Alignment—Types of Axles — 69
Wheel Alignment—Correction Methods — 72
Cam Type Adjustment for Toe — 74

CHAPTER 5

Wheel Alignment of Heavy Vehicles and Trailers — 79

Heavy Commercial Vehicles—An Overview — 79
Axle Configurations in HCVs — 80
Types of Axles in HCVs — 81
Wheel Alignment Parameters in HCVs — 83
HCV Aligner Description — 88
Infrastructure Required for HCV Alignment — 89
HCV Aligner—Field Calibration — 91
HCV Alignment—Procedure — 92
HCV Angle Corrections — 96

Chassis or Frame Alignment	103
Trailer Alignment	106

CHAPTER 6
Wheel Alignment—Points to Observe 107

Wheel Alignment—Precautions	107
Wheel Alignment—Troubleshooting	111
When to Carry Out Wheel Alignment	112

CHAPTER 7
Types of Tire Wears and the Causes 113

Tire Wear in the Middle of the Tire	113
Tire Wear in the Inner and Outer Edges	114
Tapered Wear at the Outer Edge	114
Tapered Wear at the Inner Edge	115
Feathered Wear in the Inner Edge of the Tire	115
Feathered Wear on the Outer Edge of the Tire	116
Tapered Intermittent Wear on the Inner Side of the Tire	116
Patch Type of Wear on the Tire Surface	117

CHAPTER 8
Tire Safety 119

Tire Pressure	119
Tire Rotation	121
Driver's Ability	124

CHAPTER 9
Conclusion 127

About the Author	129
Index	131

Preface

Wheels are not new to mankind. Right from the ancient bullock carts, chariots, etc., wheels have undergone many changes to serve the purpose of mankind's mobility.

Mobility is inevitable in today's life. Vehicles have become part of our day-to-day life. Without vehicles, today's world is unimaginable. An individual is willing to postpone his health checkup, but not willing to tolerate the breakdown of his vehicle. The moment a small noise or vibration is noticed, he wants to rectify it immediately. Every car owner wants to maintain his vehicle in perfect working condition. This, indeed, increases the need for dedicated vehicle maintenance and prompt services.

In a vehicle, the tire cost forms a substantial portion of the overall vehicle maintenance cost. For this reason, every vehicle owner expects his tire to last longer; in other words, he expects the tire to wear out lesser. There are many reasons for tire wear. Among them are improper Wheel Alignment and Wheel Balancing are the major reasons, which lead to severe tire wear.

Nowadays, many Wheel Alignment centers are available to offer these services. Many brands of Wheel Alignment computers are available in these centers. However, performing proper Wheel Alignment depends on the knowledge and skill of the technicians who execute the task of Wheel Alignment. Though the Wheel Alignment computer takes care of the important technical aspects of alignment, there are many mechanical activities to be carried out by the technicians during the process of Wheel Alignment to ensure proper Wheel Alignment. This is also true for Wheel Balancing. If the mechanic is not well trained, there is a possibility of him committing mistakes, leading to misalignment.

Particularly, in Multi-Axle Trucks and long Trailers, where the number of axles and wheels are many, Wheel Alignment becomes a very critical activity. Even a small misalignment will lead to severe tire wear of all the wheels, resulting in a huge loss to the vehicle owner.

This book is written for the benefit of all mechanics, technicians, and engineers who are involved in the activity of Wheel Alignment and Wheel Balancing. Training with proper knowledge goes a long way in ensuring flawless service. This book offers the required knowledge on this unique subject. Certain design aspects of Wheel Alignment and Wheel Balancing technology have also been discussed in simple language for the benefit of design engineers.

I personally feel that, besides technicians and engineers, every vehicle owner must have some knowledge on this subject. The reason being the activity of Wheel Alignment and Wheel Balancing is all about the "wellbeing" of his vehicle and more importantly his road safety.

I have used my thirty years of experience in the design, development, and patenting of some products in the field of Wheel Alignment to write this book and share knowledge.

I will be too happy if the book is going to be of any help to the mechanics, technicians, engineers, and even the vehicle owners in terms of enhancing their knowledge.

R. Mananathan
Chairman
Manatec Group of Companies
Pondicherry, India   03-08-2021

Foreword by Gurmeet Singh

I have gone through the book on "**Automobile Wheel Alignment and Wheel Balancing**," authored by Mr. R. Mananathan, Chairman, Manatec Electronics Pvt. Ltd. Initially, I thought the book is for professionals but was surprised to understand that it is a must-read for all those who own and use automobiles, irrespective of their vocation. The subject of Wheel Alignment and Wheel Balancing to which we give scant attention, I learn through the author's lucid explanations, in fact, has so much importance that leads to savings in tire cost, comfort in riding, and contribution to the environment. This book contains numerous detailed illustrations, which though simple, but with elaborate descriptions, become very clear and educative, for engineers in the automobile field.

I extend my compliments to the author for his efforts in bringing out such a useful handout for the professionals, in particular, and the public, in general. I am confident that the author continues his efforts to bring out more such books on varied subjects useful to society.

(Gurmeet Singh)

Foreword by R. Sivanesan

The book on Wheel Alignment and Wheel Balancing written by Mr. R. Mananathan is a real wealth of knowledge for everyone fascinated with Automobile Technology.

The way this book is written shows the depth of knowledge of the author. While the technology is very complex, the way the author explains every concept step by step, building from the basics, is amazing and very easy to understand. In fact, it is a delight for every Automobile Engineers to go through this book.

Mr. R. Mananathan not only explains the importance and benefits of Wheel Alignment and Wheel Balancing but also explains in a beautiful way how the Wheel Alignment and Wheel Balancing machines work, including the scientific principles of measurement.

In every chapter, he also explains what precautions to be taken and what mistakes (which are common) should be avoided.

If every workshop follows the guidelines given in this book, we can save a huge amount of money spent on tires, reduce carbon pollution, and save tons and tons of rubber, which will be a great service to the country and environment.

A must-read for every automobile enthusiast.

With best regards,
R. Sivanesan
President (Quality and AM)
Ashok Leyland Ltd., Chennai

Foreword by S. Punnaivanam

Manatec Electronics India Pvt. Ltd., led by Mr. R. Mananathan is one of the pioneers in India to provide fully indigenous Wheel Alignment and Wheel Balancing solutions for passenger cars, LCV, and HCV. Who else than Mr. R. Mananathan can be a more appropriate author for a book that explains the needs, concepts, and details of Wheel Alignment and Wheel Balancing in such a reader-friendly manner.

Hyundai Motor India has been associated with Manatec and Mr. Mananathan as an OEM recommended supplier of wheel service and automotive workshop equipment for the last 15 years. Manatec alignment and balancing equipment have always been known for their quality, durability, and cost-effectiveness for automotive workshops.

This book by Mr. R. Mananathan gives a vivid explanation of all aspects of wheel service. The book not only covers the technical aspects of alignment and balancing for the detailed understanding of machine users but also explains the importance of every process for a trouble-free and safe driving experience for vehicle owners and fleet operators.

Mr. Mananathan has written a book brings a great treasure of knowledge to the staff of automotive workshops, wheel service technicians, engineers, students, and users of automobiles. He is to be congratulated for this achievement.

I wish the author many more years of service to the automotive industry with the same zeal and passion he has today. I expected to see him continue to be a role model of "Make in India" for the Automotive workshop equipment industry.

S. Punnaivanam
VP & National Service Head
Hyundai Motor India Limited
01-08-2021

1

Wheel Alignment and Wheel Balancing

Contents

Overview 1
Vehicles and Wheels..1
Causes of Tire Wear 2
Wheel Alignment and Wheel Balancing...2

Overview

Humanity depends on the vehicles deployed for mobility. This is relevant, not only for the mobility of human beings but also for the movement of goods and materials that are involved in their day-to-day life. Life will come to a standstill if mobility is stopped.

In the past few decades, extensive research has taken place in the transport segment, and innovative products, such as the electric vehicle and driverless vehicle, have come up. Every vehicle manufacturer is brewing up new features and technologies in their vehicle in order to attract customers. In due course of time, technology will transform drastically, and it will be no wonder if we witness a day when human beings would fly.

Vehicles and Wheels

It is known that every vehicle needs wheels to run on the road, be it the most expensive vehicle or an electric vehicle or even the age-old bullock cart. To make the travel a smooth and pleasant one, the wheels are fitted with tires that are filled with air to offer the necessary cushioning effect. These air-filled tires bear the entire load of the vehicle and also absorb the varying forces and vibrations that arise during the movement. Since tires are the interface between the road and the vehicle, they withstand these forces and, hence, are subjected to abnormal friction. This friction causes the tires to wear out. Consequently, these worn-out particles of the tires get dissolved in the air in the form of 'particulate matter' and increase the pollution levels in the air.

Causes of Tire Wear

- Excess or low air pressure in the tires.
- Improper wheel alignment and wheel balancing in the vehicle.
- Rough roads.
- Frequent application of brake by the driver.
- Poor quality of tires.

If all the above causes are eliminated, a tire can last for more than a hundred-thousand-kilometers (62,100 miles) run before it wears out fully.

Among the above-listed causes, wheel alignment and wheel balancing are considered to be the most important causes of tire wear. Although the owner of the vehicle can control and mitigate the other causes of tire wear, he must depend on the services of a technician for the wheel alignment and wheel balancing of his vehicle. Therefore, it is advisable that the vehicle owner also learns the basic principles of wheel alignment and wheel balancing.

Wheel Alignment and Wheel Balancing

Wheel alignment and wheel balancing are two different subjects and involve two different activities. Wheel balancing must be done first, and balanced wheels must be fitted to the vehicle prior to Wheel Alignment.

Wheel Balancing

Wheel Balancing must be done for every wheel of the vehicle. It is the activity of removing any unbalance in a particular wheel and making it a balanced one. If wheel balancing is not proper in a vehicle, the following problems will be faced:

1. The vehicle will travel with vibrations.
2. Patch type of wear can be seen on the peripheral surface of the tires. These patches will soon tear the tires.
3. All the suspension parts of the vehicle will get rattled and affected.
4. Driving will be uncomfortable. Vibrations can be felt in the steering wheel.

Wheel Alignment

Wheel Alignment is an act of making all the four wheels of a vehicle travel in a unified direction. This is achieved by maintaining the wheel alignment angles such as Camber,

Toe, Caster, and King Pin as per the specifications of the vehicles. In other words, the activity of wheel alignment can be defined as below.

> The process of measuring and ensuring the various wheel alignment angles specified by the manufacturer in all the wheels of a vehicle, both individually and collectively, to achieve unidirectional running, is called wheel alignment.

Nowadays, this is done with the help of wheel alignment computers. These computers measure the various wheel alignment angles and tell us which angles are within limits and which angles are not. The angles that are not within limits will be corrected during the wheel alignment process and made alright for the unidirectional running of all the wheels in the vehicle.

If wheel alignment is incorrect in a vehicle, the following problems will be encountered:

- The vehicle may not travel straight. It may get dragged to one side.
- The steering wheel may be squinted instead of looking straight.
- The tires will be subjected to abnormal wear and tear, resulting in decreased tire life.
- When the brake is applied at high speeds, the vehicle may get dragged to one side before halting, thus leading to accidents.
- Increased fuel consumption.

If every vehicle in a country is maintained with proper wheel alignment and wheel balancing, there will be a huge saving in tire expenditure, fuel efficiency, and road safety. Also pollution levels will decrease, thus indirectly protecting the environment.

Assuming that a tire manufacturer produces high-quality tires, but fails to ensure proper alignment and balancing in the vehicles, then the tires will wear out quickly, tarnishing the reputation of the tire manufacturer. For this reason, all the tire manufacturers include alignment and balancing in their dealership points to ensure compliance with these activities. This is essential for the survival of the tire manufacturer in a competitive marketplace.

This is equally important for the vehicle manufacturers also because when a customer buys a vehicle, his expectations are smooth running and minimum maintenance. All the vehicle dealers also have alignment and balancing services in their facilities.

In the coming chapters, we will learn more about Wheel Alignment and Wheel Balancing.

2

Wheel Balancing

Contents

What Is Wheel Balancing? 5
Reasons for Unbalance in Wheels..7
Single Plane Balancing...10
Two Plane Balancing..12
Wheel Balancing Machine—Description..13
Wheel Balancing Machine—Installation 14
Wheel Balancing—Procedure..14
Wheel Balancing—HCVs...17
Wheel Balancing—Two Wheelers...18
Wheel Balancing—Important Points..19

What Is Wheel Balancing?

We all know that the wheel of an automobile is circular in shape. This circular wheel is fitted to the axle of a vehicle through a bearing, and when the vehicle moves, the wheel rolls on the road. While rolling, do you know what will be the rotational speed of a wheel?

The rotational speed of any object is measured in **revolution per minute** (rpm). For the same vehicle speed, when the diameter of a wheel is less, the rpm will be high. This can be understood from the following chart (Table 2.1).

TABLE 2.1 Relationship between speed, wheel diameter, and rpm.

Vehicle speed (km/hour)	120	120
Wheel diameter (cm)	55	35
Wheel speed (rpm)	1160	1820

A wheel with 55 cm diameter, at a vehicle speed of 120 km/hour, will rotate 1160 times in a minute (rpm). Whereas a wheel of lesser diameter, say, 35 cm, will rotate 1820 times for the same vehicle speed of 120 km/hour. In both cases, the rpm is very high. In such high rpm, even a small unbalance in the wheel will cause undesirable effects on the wheel, and the tires are susceptible to damages. Therefore it is important that we know more about the **unbalance** and the forces created by such unbalances in automobile wheels.

FIGURE 2.1 Illustration of centrifugal force.

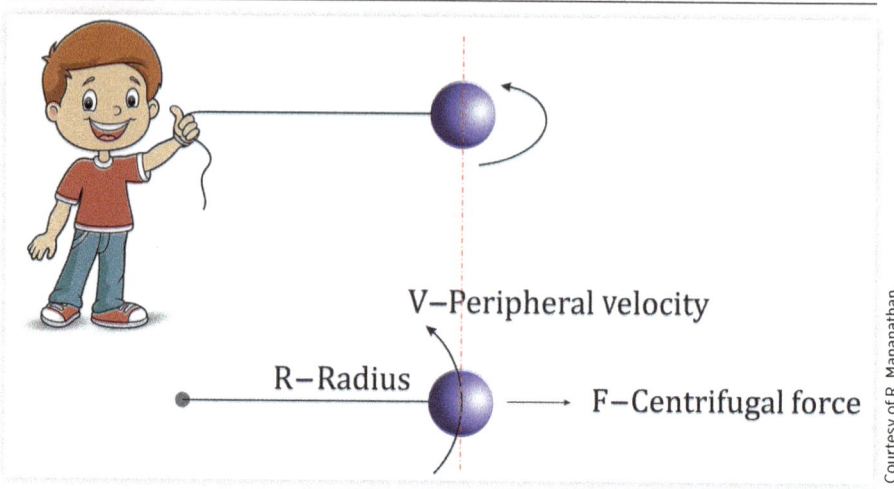

Look at Figure 2.1. Assume that a boy ties a ball to a string and swings it around himself as shown in the picture. If seen from the top, the ball will describe a circle around the boy and the ball will try to fly away from his hand. But the boy tightly holds the ball to prevent it from leaving his hand. The force that tries to pull the ball away from his hand is called **centrifugal force**. The value of this centrifugal force can be calculated using the following formula.

$$F = \frac{MV^2}{R}$$

where
    F is the centrifugal force
    M is mass of the ball
    V is the peripheral velocity
    R is the radius

From the above formula, it can be understood that, when the rpm increases, the peripheral velocity **V** also increases. This means **V** is directly proportional to rpm. If the speed **V** increases, the centrifugal force **F** increases in square proportion. In other words, if **V** increases by two times, the centrifugal force will increase by four times, and if **V** increases by three times, the centrifugal force will increase by nine times. This means the centrifugal force F is proportional to Velocity^2 or F is proportional to rpm^2 (F α V^2 or F α rpm^2)

From the above, it can be understood that, **if there is an unbalanced mass in a rotating object, it results in a centrifugal force. This centrifugal force increases in proportion to the square of the speed.**

Since automobile wheels are running at high speeds, even small unbalances in the wheels will lead to high centrifugal forces while running on the road. These centrifugal forces will squeeze the tires in each and every rotation and cause damage to the tires.

Reasons for Unbalance in Wheels

A wheel is an assembly of a **rim** (made of steel or aluminum alloy) and a tire (made of rubber). Both the rim and tire are circular in shape. Since they are **circular** in shape, the mass around the center point of the wheel must be distributed uniformly on the periphery, both in the wheel rim and in the tire. But, in practice, it may not be uniform in all the wheels. Even in new wheels, the mass distribution will not be uniform.

The steel wheel rims are manufactured through different processes like forming, pressing, welding, etc. The aluminum alloy rims are mostly pressure die casted or forged. In these manufacturing processes, there are possibilities for the mass distribution to be not uniform around its center point. During the process of casting, air bubbles may get trapped. Due to these reasons, even new rims may have a certain unbalance.

In the case of tires, they are manufactured through a molding process. While molding the tires, the molten rubber is supposed to flow and spread uniformly around the periphery. But, in practice, the temperature of the molten material may not be uniform throughout the process, and during cooling, in certain spots the density of rubber may vary. In such spots, the **mass** will be either more or less compared to the other areas in the periphery. Secondly, air bubbles may also get created during the molding process in spite of the care taken during the process of molding. Such low-density spots or air bubbles can cause unbalance in the tires.

In the case of old vehicles, while running on the roads, uneven tire wear happens, which affects the wheel balance. When the vehicle runs on rough roads, it may hit some potholes, ridges, etc., subjecting the wheels to abnormal forces, which will distort the wheel rim. This also affects the wheel balance.

In short, the unbalance in wheels can be due to any one of the following reasons:

New wheels

- Wheel rim distortion during manufacturing.
- Low- or high-density spots in tires during the molding process.

Old wheels

- Uneven tire wear while running on roads.
- Wheel rim distortion due to rough roads.

FIGURE 2.2 Excess mass and unbalanced force.

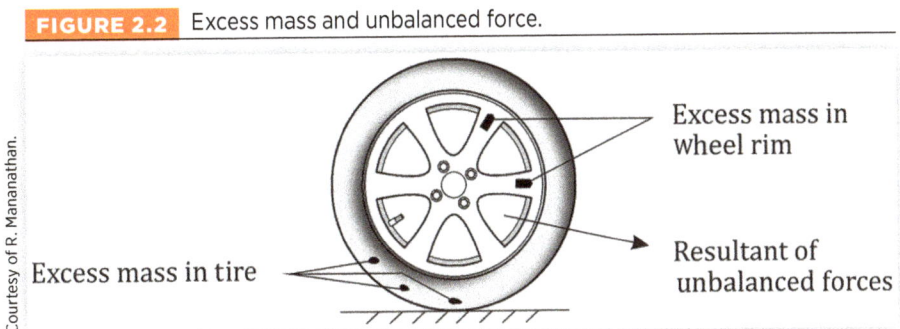

When the vehicle runs on the road, the above unbalances in a wheel can cause centrifugal forces in many places, depending on the existence of extra masses and their locations.

In practice, we will not know from which spot(s) on the rim or tire such centrifugal forces get developed. But all these forces put together generate a **resultant force** while running which acts on a radial line emanating from the center point of the wheel.

FIGURE 2.3 Resultant centrifugal force touching the road.

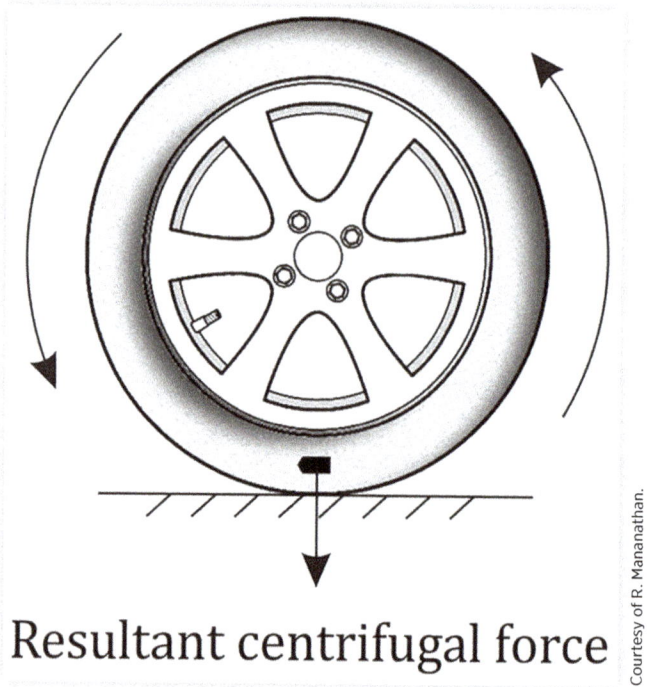

When this resultant centrifugal force touches the road during a rotation as shown in Figure 2.3, an equal and opposite force is generated at this point that squeezes the tire. The moment this point leaves the touching spot on the road, the squeezing is released. This effect of **squeezing and releasing** happens in every rotation of the wheel. When the rpm increases, this effect not only increases the frequency of such **squeezing and releasing** but also increases the centrifugal force in square proportion to the rpm, thereby increasing the severity of the **squeeze**. This squeezing action results in a patch type of wear at this point on the tire.

This **squeezing and releasing** phenomena makes the driver to feel a vibration on the steering wheel while driving. At high speeds, the vibration will be more and the driver will feel chattering in the steering wheel. Also the ball joints, kingpin bushings, and other joints of the vehicle will be subjected to unwanted forces. In short, unbalance in wheels will rattle the vehicle.

Such wheels having extra masses (weights) are called **unbalanced wheels**. Wheel Balancing is the process by which these wheels are made **balanced**.

To make these wheels a balanced one, the extra weight (mass) has to be identified and removed. But practically, it is not possible to remove the extra weight from the rim or tire. Instead, we can add equal weight on the opposite side of the wheel to create a centrifugal force opposite to the centrifugal force created by the unbalanced mass. This will nullify the effect of unbalanced centrifugal force and will make the wheel a **balanced** one. This principle is followed in all Wheel Balancing machines.

If the wheels are left unbalanced, the wheel and the vehicle will be subjected to the following damages.

On the road, where the unbalanced force touches and leaves, the tire will undergo a spot wear or a patch type of wear as shown in Figure 2.4.

FIGURE 2.4 Patch-type wear.

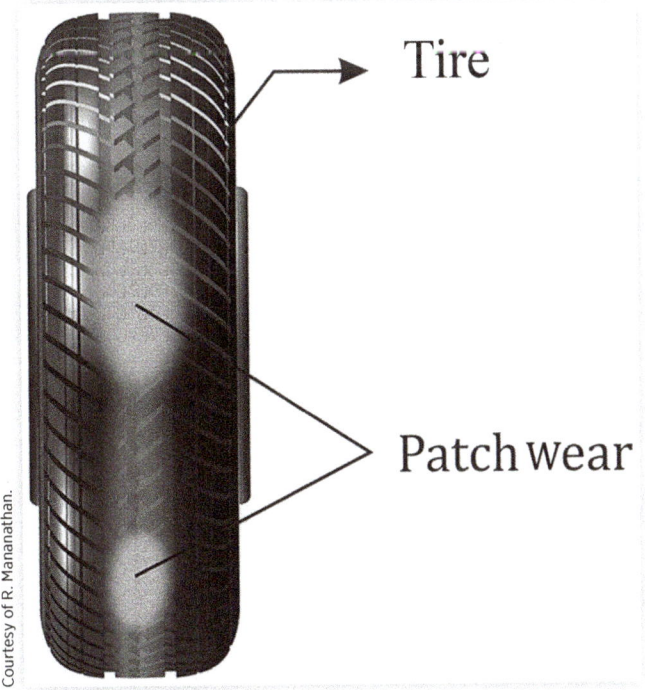

Courtesy of R. Mananathan.

This type of wear will soon tear the tire and damage it. The vehicle will vibrate and rattle, damaging many parts of the vehicle. The drive will not be pleasant. All the above problems can be prevented if Wheel Balancing is done periodically. Also the tire life will increase and the **drive** will be pleasant. Now we will understand how the **Wheel Balancing** works.

Single Plane Balancing

Let us assume that some unbalanced mass exists in a wheel in point A as shown in Figure 2.5. The wheel is fitted to a shaft rotating on bearings.

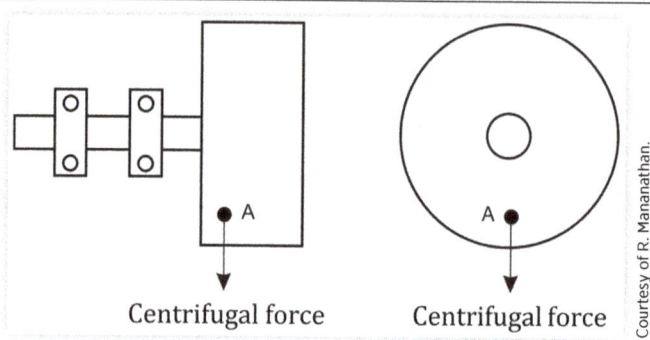

FIGURE 2.5 Effect of unbalance.

If the wheel is rotated by hand and left free to stop, being an unbalanced wheel, the heavy point A comes down to the bottom and stops. This happens every time the wheel is rotated and left free to stop.

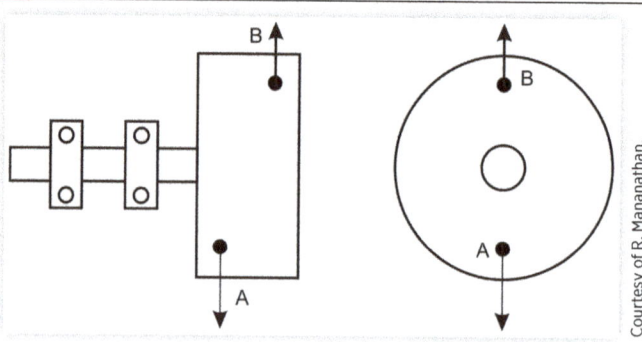

FIGURE 2.6 Single Plane Balancing.

Now, if a weight equal to mass A is added in location B, which is on the opposite side of location A, as shown in Figure 2.6, the wheel gets balanced. If the wheel is rotated and left free to stop, the same point will not come to the bottom position. Every time this is done, the wheel stops in different positions. This means the wheel got balanced. This method of balancing is called **Single Plane Balancing**. This is also called **Static Balancing**.

But there is a drawback in this method.

FIGURE 2.7 Force couple.

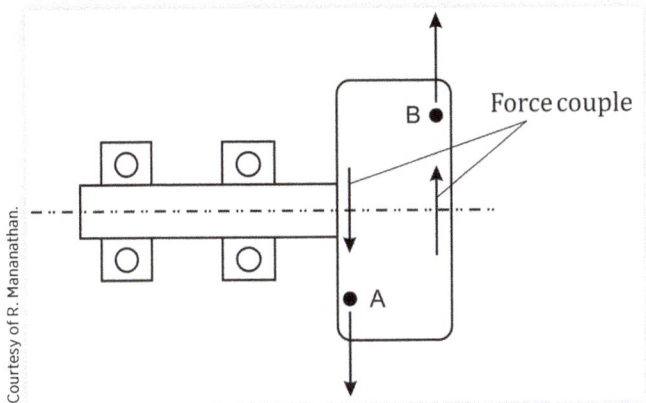

As seen in Figure 2.7, though the weight is added on the opposite side in point B, it is not located in the same plane of point A. Because of this, in spite of achieving the **balance**, a centrifugal force in plane A and a centrifugal force in plane B get generated separately, resulting in a Force Couple. This Force Couple will act on the bearings, exerting pressure, which will damage the bearings over a period of time.

FIGURE 2.8 Ideal location for weight addition.

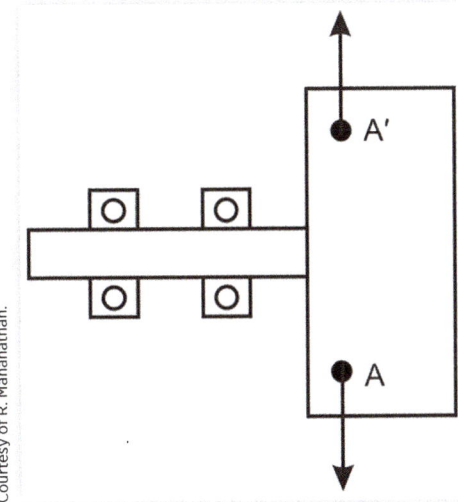

For this, the solution will be to add the weight in Point A' which is exactly at 180° opposite to point A in the same plane as shown in Figure 2.8.

Now the wheel will get balanced and the force couple will not get generated. In practice it is not possible to know the exact spot of point **A** and its plane. Point **A** may be anywhere along the width of the wheel, which is impossible to locate. Also it may be at any radius.

Under these circumstances, a method called **Two Plane Balancing** will be useful.

Two Plane Balancing

FIGURE 2.9 Two plane resolving.

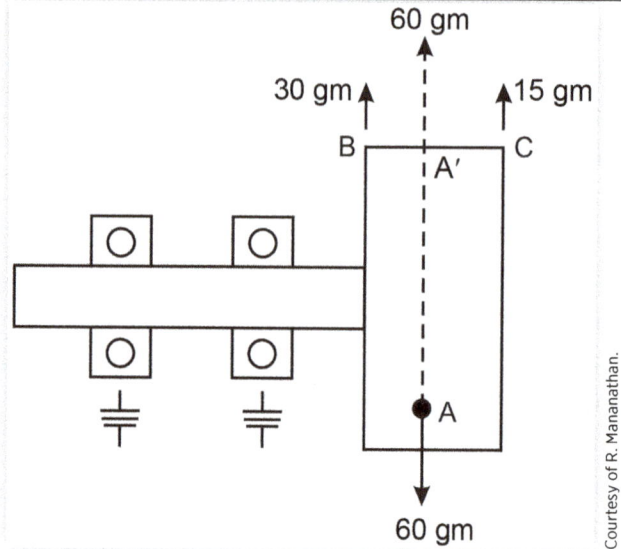

Let us assume that an unbalance of 60 grams exists in plane A as shown in Figure 2.9. To make the wheel balanced without a Force Couple, we must add 60 grams weight in the same plane of **A** and at 180° opposite to point **A**. But plane **A** is not known to us. To resolve this issue, the following method is followed.

FIGURE 2.10 Weight calculation.

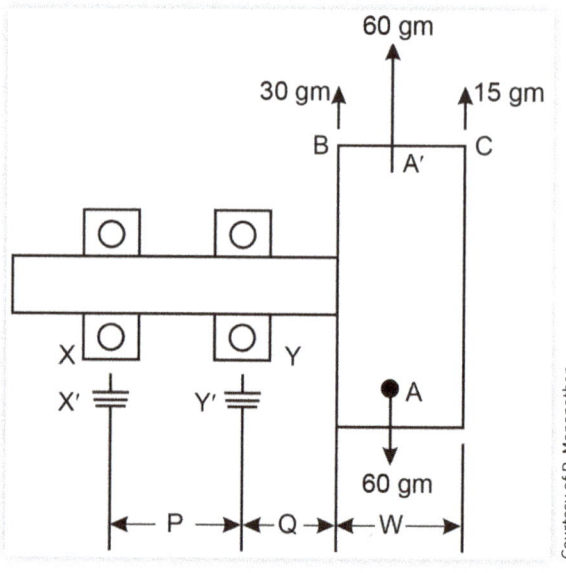

As seen in Figure 2.10, the unbalanced wheel is fitted to a shaft and the shaft is fitted with two bearings X and Y. In each bearing, a **Piezo Electric Sensor** is fitted. Let these sensors be called X' and Y'. When the wheel rotates, the centrifugal force generated in point A is acting on the two bearings X and Y in proportion to the distance of the bearings from plane A. Consequently, these forces are transmitted from the bearings to the sensors X' and Y'. These forces will be repeating in every rotation. Therefore, they create a vibration on the sensors with a frequency equal to the **rpm**. The amplitude of the vibration will be proportional to the forces acting on the bearings.

These vibrations acting on the Piezo Electric Sensors can be measured and converted into forces. Using these forces and also using the distances P, Q, and W, we can resolve the forces in planes B and C, by applying a suitable mathematical formula. These forces in planes B and C can be converted into weights using the wheel rim diameter (D) where it is convenient to add weight.

The Wheel Balancing machines perform this activity and give output of how much weight has to be added in planes B and C, which are the inner and outer rims, respectively.

Therefore, **Two Plane Balancing** can be defined as the act of resolving an unknown force (A), into forces at two known planes (B and C), in such a way that the resultant of these two forces are equal to force A and acts in the desired plane (A') in the direction required for balancing.

This method is also called **Dynamic Balancing**, and universally all Wheel Balancers work on this principle only.

Wheel Balancing Machine—Description

A Wheel Balancing machine will have the following parts as shown in Figure 2.11:

FIGURE 2.11 Wheel balancing machine.

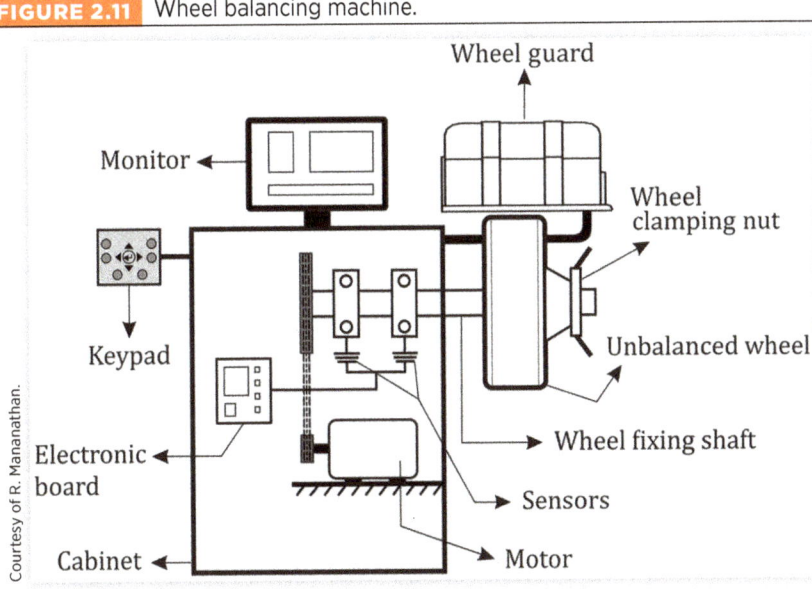

1. **Wheel Fixing Shaft:** Designed to rotate on two bearings with suitable threaded ends to fix the unbalanced wheels.
2. **Wheel Clamping Nut:** Used to fix and clamp the unbalanced wheel on the shaft and lock it firmly on the shaft end.
3. **Motor:** Used for rotating the shaft through a flexible belt.
4. **Keypad:** Alphanumeric Keypad to input the distance parameters.
5. **Sensors:** Deployed for sensing the unbalanced forces and sending them to the Electronic Board for further processing.
6. **Electronic Board:** Receives the signals from Sensors and processes the signals to give the output in terms of weight to be added in the inner and outer rims (planes) of the wheel. (A computer can also be used instead of the electronic board to process the data and show the results pictorially in a display unit.)
7. **Display:** This can be a simple Seven Segment Display or a computer Monitor, depending on the model designed.
8. **Cabinet:** A metal box designed to accommodate all these parts.
9. **Accessories:** Wheel diameter measuring caliper, weight fixing hammer, etc. are the accessories needed for a Wheel Balancer.

Wheel Balancing Machine—Installation

Normally, it is not necessary to firmly fix a Wheel Balancer on the floor using fixing bolts. But the balancer must be kept on a levelled floor. The balancer can be fixed using bolts on the floor for preventing the balancer from moving away while fixing the weights using a hammer. If the Wheel Balancer is tightened using fixing bolts on an uneven floor, the cabinet will get strained, and consequently, other parts may also get strained. This kind of strained fixing will make the Wheel Balancer to give wrong results. Therefore, while installing a Wheel Balancer, strain to the cabinet must be avoided.

However, it is advisable to follow the procedure given in the User Manual given by the manufacturer. During installation, the balancer shaft must not be used for lifting or moving the machine. If the shaft is used for moving the machine, it is likely to get bent and will seriously affect the balancer's performance.

Wheel Balancing—Procedure

Fix the unbalanced wheel onto the shaft extending out of the Wheel Balancing machine and firmly tighten the same using the Wheel Clamping Nut supplied along with the machine.

Input the distances P, Q, W, and D as per Figure 2.10 into the system using the Alphanumeric Keypad available in the machine.

Close the wheel using the wheel guard.

Now press the **Start** button. The motor in the machine will rotate the wheel at a low rpm as per the design.

This rotation will stop after a few seconds, and during this short period of rotation, the Piezo Electric Sensors fitted to the bearings acquire the vibrations created by the unbalance and transmit to the Electronic Board in the machine. This Electronic Board contains a microcontroller, which receives the above data. This microcontroller already has the data of the distances, diameter, etc., which were keyed in as input. Using this data and the sensor outputs, the microcontroller calculates the weight to be added to the inner and outer rims of the wheel. These are displayed in the display as shown in Figure 2.12.

FIGURE 2.12 Results display.

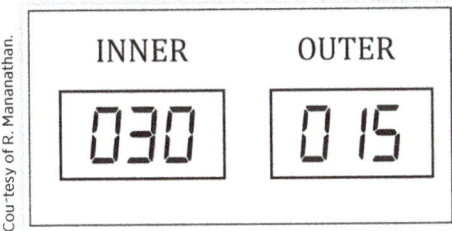

The display shows that 30-gram weight is to be added onto the **inner rim** and 15-gram weight to be added onto the **outer rim**. Now these weights have to be added to the rims as per the displayed results. But the inner and outer rims are circular in shape, and the following question arises.

In the 360° of the circle, where do we add the weight?

To locate the exact place where the weight has to be added, a picture is given in the machine as shown in Figure 2.13. This figure shows both the inner and outer planes. As shown in the figure, keep rotating the wheel slowly by hand. The moment the point where the weight has to be added in the inner wheel rim comes to the top position, a light-emitting diode (LED) light on the inner side will glow. Stop at this point and fix a 30-gram weight at the top point on the inner wheel rim. While fixing the weight, press the brake pedal using the leg to prevent the wheel from rotating.

FIGURE 2.13 Phase location.

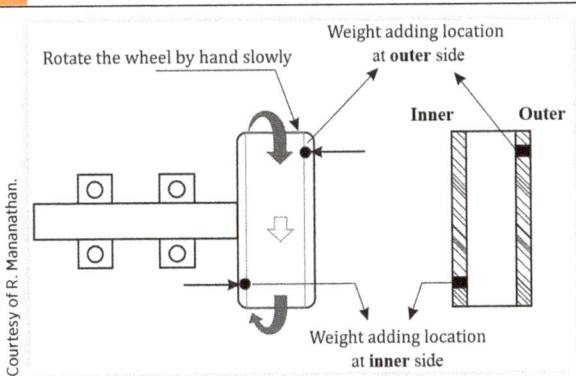

Similarly, if the wheel is slowly rotated further, the moment the point where the weight must be added on the outer wheel rim comes to the top position, an LED light will glow. Stop at this point and fix the 15-gram weight at the top point on the outer wheel rim. While fixing the weight, press the brake pedal to prevent the wheel from rotating.

FIGURE 2.14 Wheel balancing weight.

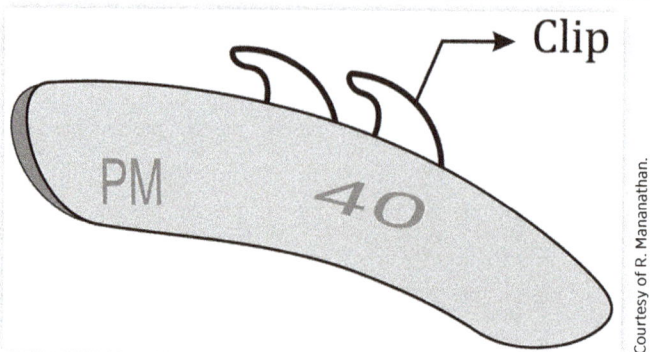

Figure 2.14 shows a hook in the weight to prevent the weight from coming out after fixing. These hooks are made of spring steel, which has the necessary flexibility for firmly gripping the rim after fixing.

Wheel Balancing weights ranging from 5 grams to 200 grams, and even more for heavy commercial vehicles (HCVs), are available in different weights in the market. These weights have to be fixed onto the rim using the weight fixing hammer supplied along with the machine. The steel clip may damage the alloy wheel rims during hammering and fixing. To avoid this, **sticker** weights are available in the market as shown in Figure 2.15.

FIGURE 2.15 Sticker weight.

These sticker weights are flat in shape. These are flexible in such a way to enable pasting in the inner side of the rims. These sticker weights have **glue** at one side for easy pasting.

After fixing the correct weight in the respective planes, to find out whether the balancing has been achieved, again press the **Start** button and run the balancer. Now the display will show zero values, both in the inner plane and in the outer plane (Figure 2.16).

FIGURE 2.16 Final results display.

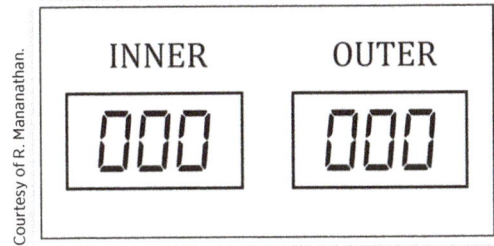

If zero is displayed for both the planes, it means the balancing of the wheel is completed. Instead of zero if some values are displayed, we have to fix the weights as per the values shown in the display and again run the balancing program by pressing the **Start** button. Now the balancer is supposed to display **zero** values.

If not, this process must be continued till **zero** value is achieved in both planes:

 Note: Normally the wheel will get balanced in one run itself. In some cases, it may require two runs. If the balancer requires more than three runs, the possible reasons are listed below. The correct reason must be found and corrected.

- The wheel might have not been fitted properly on the shaft (looseness or eccentric fitting).
- Due to an accident, there may be more distortions in the wheel rims.
- The balancer may have some electronic problem.

If any one of the above problems exist, it must be attended and set right before using the balancer further.

Wheel Balancing—HCVs

Since the wheel diameter of the HCVs is more than the wheel diameter of light commercial vehicles (LCVs), the rpm of the HCV wheels will generally be less for the same speed. Consequently, the centrifugal forces will be less. For this reason, there is a general feeling that Wheel Balancing is not necessary for HCVs. This is **not** correct.

Though the rpm of HCV wheels are low, the unbalanced **mass** in HCVs are generally much higher than LCVs resulting in more centrifugal force. Also, with improved road conditions, the average speed of HCVs has gone up. This also causes higher centrifugal force. Therefore, in HCVs too, balancing of the wheels is very important to reduce tire wear and to increase the tire life.

FIGURE 2.17 HCV wheel balancer.

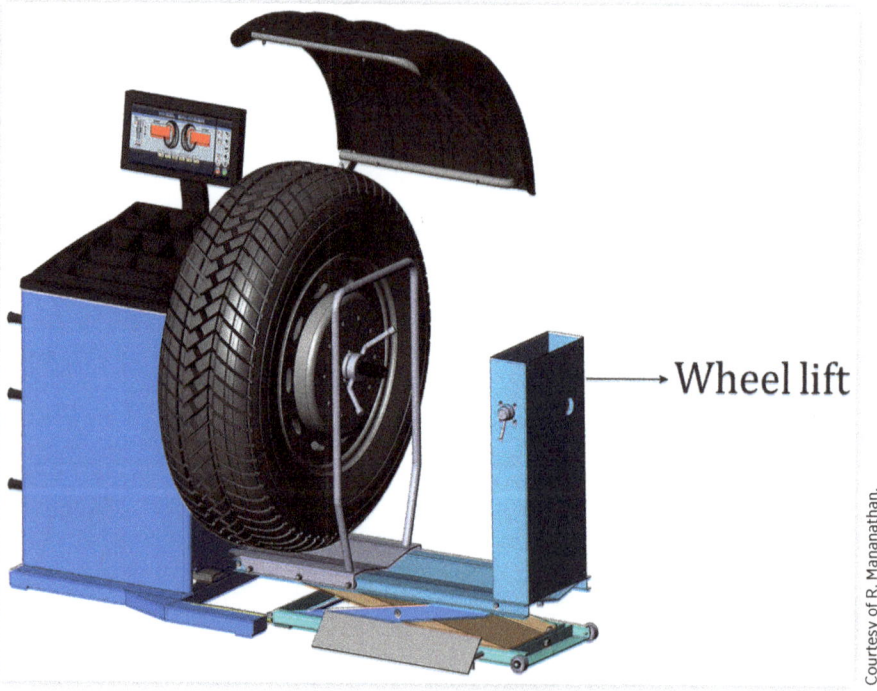

The technology of Wheel Balancing is common for both LCVs and HCVs. Since the HCV wheels are bigger and heavier than LCVs, the design of balancers for HCVs must be rugged and strong enough to withstand the heavy weights and the centrifugal forces. Accordingly, the design of motors, shaft, sensors, bearings, etc. will be higher in size than LCV balancers. As shown in Figure 2.17, a lift is provided to help the technicians to lift the heavy wheels and fix it onto the Wheel Balancer shaft. For these reasons, the HCV balancers are more expensive than LCV balancers.

The operating procedure of an HCV balancer is the same as the operating procedure of an LCV balancer.

Wheel Balancing—Two Wheelers

Generally, the Two Wheelers travel at high speeds and hence, the centrifugal forces arising out of even small unbalance will be very high. For this reason, Wheel Balancing is important for two wheelers as well. When these wheels are not balanced properly, they face dynamic **instability** on the roads. Normally, in two wheelers, the wheels will have some amount of **Run-out** and **Face out**, which are undesirable. These can be measured and set right by using Wheel Balancers having this facility. Besides reduction in tire wear, preventing possible accidents at high speeds is the most important benefit of Wheel Balancing in two wheelers:

Wheel Balancing—Important Points

1. Unbalance will be present in new wheels as well. This is because of certain inherent defects that occur during the production of wheel rims and tires.
2. Even a perfectly balanced wheel becomes unbalanced due to uneven tire wear during running.
3. While running on roads, if the vehicle hits a pothole, the rim may get distorted resulting in wheel unbalance.

Therefore, it is recommended to carry out Wheel Balancing every 10,000 km running of the vehicle. If the vehicle is going to operate on rough roads, Wheel Balancing is recommended every 5000 km.

3

Wheel Alignment of Light Vehicles

Contents

Light Vehicles—Definition 21
Wheel Alignment Angles and Parameters 21
Parameters affecting Wheel Alignment 34
Wheel Alignment angles measuring technology and Its advancement 42
Charge Coupled Device (CCD) Technology 43
3D Technology 46
3D Wheel Alignment Computer and Its Accessories 50
3D Wheel Aligner—Installation 56
3D Wheel Aligner—Calibration 59
3D Wheel Alignment Procedure 62

Light Vehicles—Definition

Passenger Cars, Vans, and LCVs, excluding buses, trucks, and trailers are called **Light Vehicles** from Wheel Alignment point of view.

Wheel Alignment Angles and Parameters

Wheel Alignment is the process carried out after balancing all the four wheels and mounting them to the vehicle. When all the four wheels are mounted to the vehicle, it must run on the road in a straight line with **dynamic stability**. For this, the vehicle manufacturer has specified certain angles at which each wheel has to be mounted onto the vehicle axles. These specified angles are called Wheel Alignment Specifications. If the wheels are mounted conforming to these specifications, the vehicle will travel on the roads without any vibration and with less friction. The following are the Wheel Alignment angles for Light Vehicles:

1. Camber.
2. Toe and Total Toe.
3. Kingpin angle.
4. Caster.

All the above four parameters are angles measured in degrees and minutes. Among these angles, the Camber and Toe are the angles pertaining to the wheel position and therefore called **Wheel Angles**. The Caster and Kingpin angles pertain to the angle of the **Steering Axis**, which is the axis located in front of the vehicle and acts as a fulcrum for the front wheels to turn during steering. These are called **Steering Axis Angles**.

The above four angles are considered to be very important angles for Wheel Alignment. Apart from the above four angles, the following angles/parameters also pertain to Wheel Alignment geometry:

1. Included Angle.
2. Toe Out on Turns.
3. Lock Angle.
4. Ride Height.
5. Track Width and Wheel Base.

Now let us understand all the above angles and parameters in detail.

Camber

When we view the front wheels of a vehicle standing in front of the vehicle, though the wheels seem to be standing perpendicular to the road, in reality they are not. They will be inclined to the **true vertical** position either towards the inside or towards the outside of the vehicle. This inclination is called Camber.

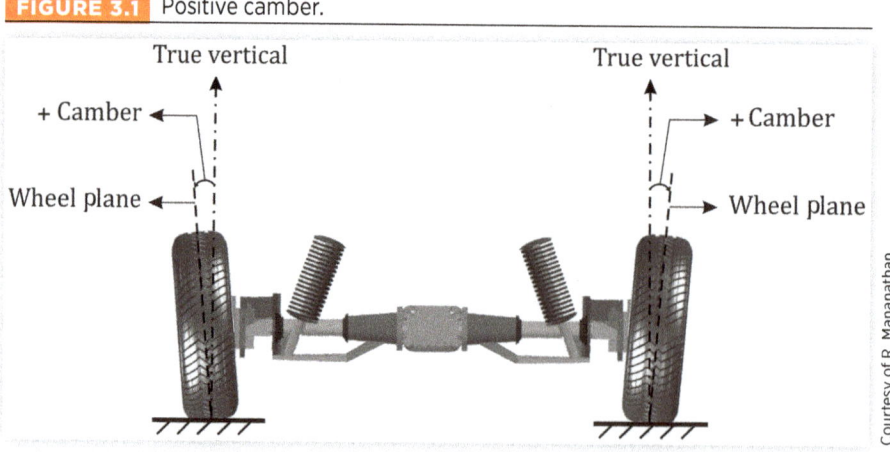

FIGURE 3.1 Positive camber.

Courtesy of R. Mananathan.

FIGURE 3.2 Negative camber.

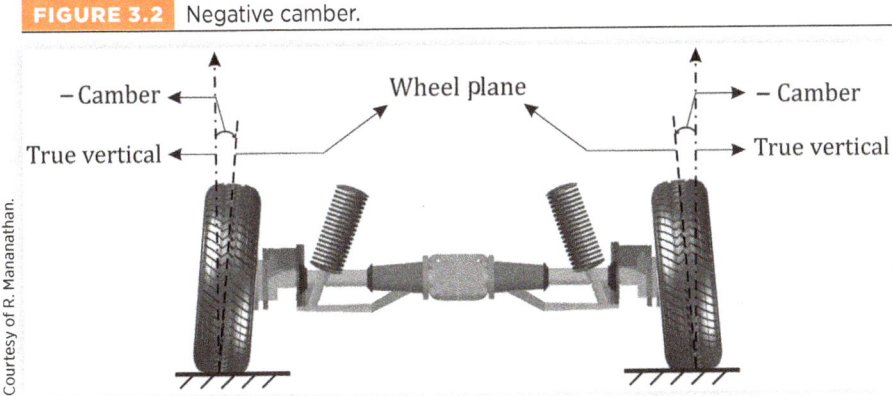

As seen in Figures 3.1 and 3.2, the inclination towards the outside of the vehicle is called Positive Camber and the inclination towards inside the vehicle is called Negative Camber. For all vehicle models, the specification of the Camber is decided by the vehicle manufacturer based on the weight of the vehicle and the maximum number of persons proposed to be travelling. Therefore the Camber can be called a **Load Bearing** angle. If the Camber is not set as per the specification in a vehicle, it will lead to tapered tire wear either on the inner side or on the outer side of the wheels as shown in Figures 3.3 and 3.4.

FIGURE 3.3 Tapered flat wear on the outer edge of the tire.

Cause: Excess **Positive Camber** or less **Negative Camber**

FIGURE 3.4 Tapered flat wear on the inner edge of the tire.

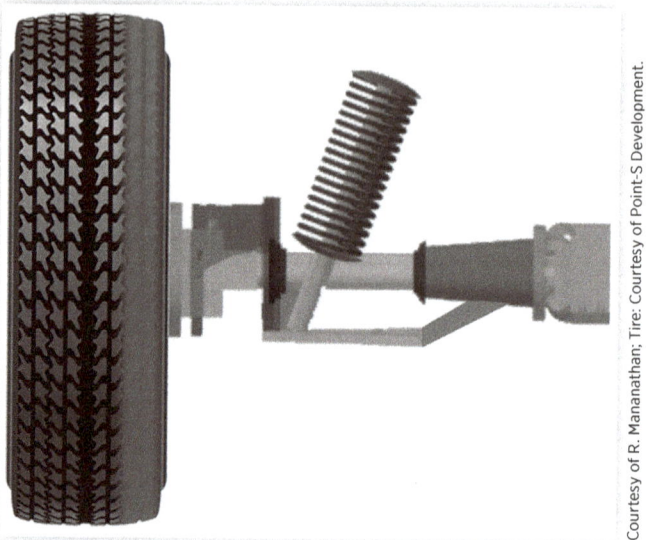

Cause: Less **Positive Camber** or excess **Negative Camber**

Toe and Total Toe

If a vehicle is viewed from its top, the left and right wheels of both the front and rear sides will be seen inclined either towards the inside of the vehicle or towards the outside of the vehicle. This inclination with respect to the Geometric Centerline (GCL) is called **Toe** angle.

FIGURE 3.5 Toe-in (Positive Toe).

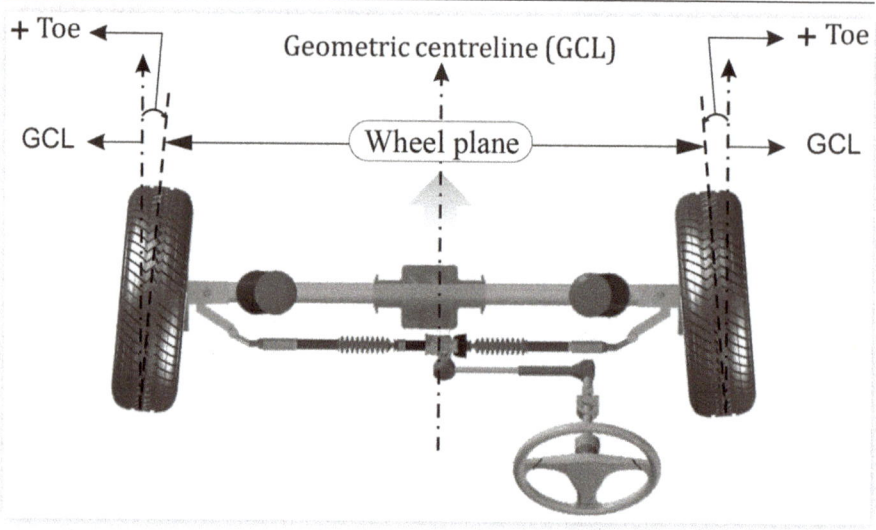

As shown in **Figures 3.5** and **3.6**, if the inclination is towards the inside, it is called **Positive (+) Toe** and if the inclination is towards the outside, it is called **Negative (−) Toe**. The Positive Toe is also called **Toe-in**. Similarly, the negative Toe is called **Toe-out**.

FIGURE 3.6 Toe-out (Negative Toe).

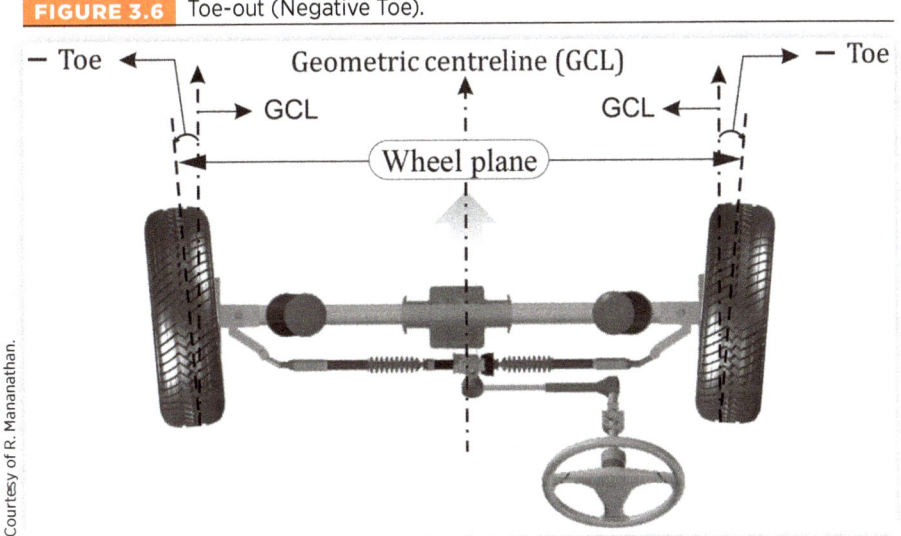

In any vehicle, the Left Toe and the Right Toe angles will be equal in value and also in sign. This means if the left Toe is **positive**, the right Toe will also be **positive** and vice versa.

Total Toe

Total Toe is the sum of the Left Toe and the Right Toe

$$\text{Total Toe} = \text{Left Toe} + \text{Right Toe}$$

Caution: In many vehicles, only the **Total Toe** is given as the specification. Considering this, if the Total Toe is set as per specification, ignoring to maintain the individual Left and Right Toe values equally, it will result in **Steering Cross**. Therefore, while setting the **Total Toe**, it must be ensured that the individual Left and Right Toe values are also set equal at half the value of the Total Toe.

The vehicle manufacturer determines the Toe specification considering the **tractional friction** and the **Dynamic Stability** of the vehicle during its motion on the road. If the Toe is not as per specification in a vehicle, the tires will wear out rapidly.

For this reason, the Toe is considered a very important and critical angle of Wheel Alignment.

If the Toe is not within limits, it will result in the following bad effects:
- Abnormal tire wear.
- Vehicle pulling to one side while driving.
- At high speeds, if the brake is applied suddenly, the vehicle, instead of stopping in the same line of travel, will get dragged to one side and then stop. This may lead to accidents.
- Steering Cross (crooked steering) after alignment, as shown in Figure 3.7.

FIGURE 3.7 Steering Cross.

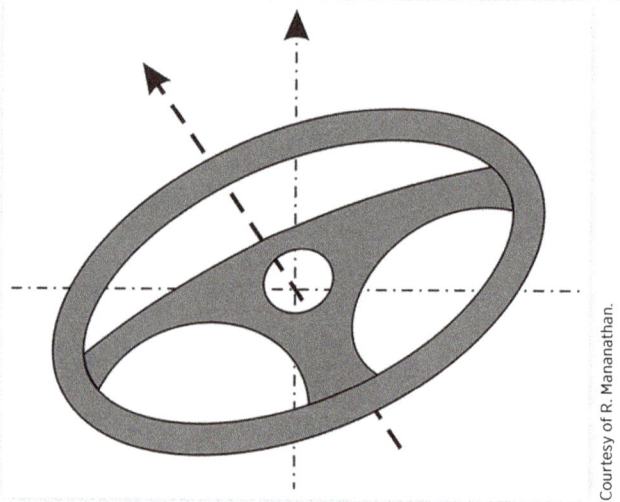

Improper Toe leads to feathered wear at the edges of tires as shown in Figures 3.8 and 3.9.

FIGURE 3.8 Feathered wear on the inner edge of the tire.

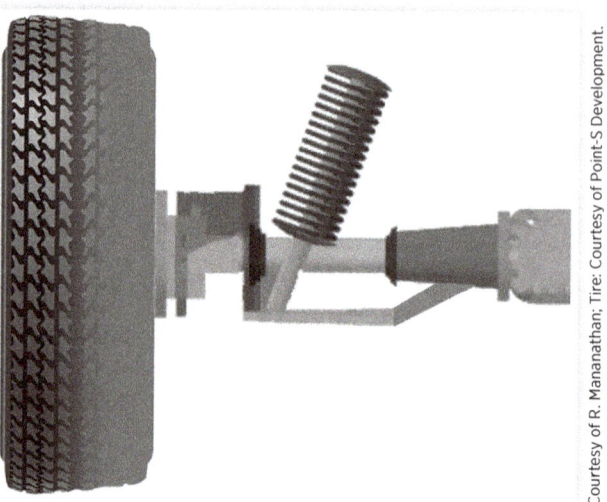

Reason: Excess **Toe-in** or less **Toe-out**

FIGURE 3.9 Feathered wear on the outer edge of the tire.

Reason: Less **Toe-in** or excess **Toe-out**

In a vehicle, if the Toe is not within specification, it leads to abnormal and severe tire wear. Therefore, extreme care must be taken to ensure that the Toe is set within the specification in a vehicle.

Kingpin Angle

When the steering wheel of a vehicle is turned, the front wheels turn using an axis as a pivot. This pivotal axis is called Steering Axis. In practice, it is also called Kingpin. There is a Steering Axis on the left side and also on the right side fitted on the Front Axle. In each Steering Axis, there are two bush bearings, one at the bottom and the other on top of the axis. The wheels are mounted to these bushes, and hence these bush bearings enable easy turning of the wheels while steering the wheels. When viewed from the front of the vehicle, both the left and right steering axis will be seen inclined towards the inside as shown in Figure 3.10. This inclination with respect to **true vertical** is called the Kingpin Angle. The Kingpin angle is always inclined towards the inside of the vehicle. Kingpin angle is also called **Steering Axis angle**.

Kingpin angle is a load bearing angle. It guides the vehicle for smooth traction, particularly on curves. The Kingpin angle is a fixed angle and cannot be corrected during Wheel Alignment. But this angle is measured during Wheel Alignment. If the Kingpin angle is deviating from the specification, it can be corrected only in the workshop and not in the Wheel Alignment Center.

FIGURE 3.10 Kingpin angle and scrub radius.

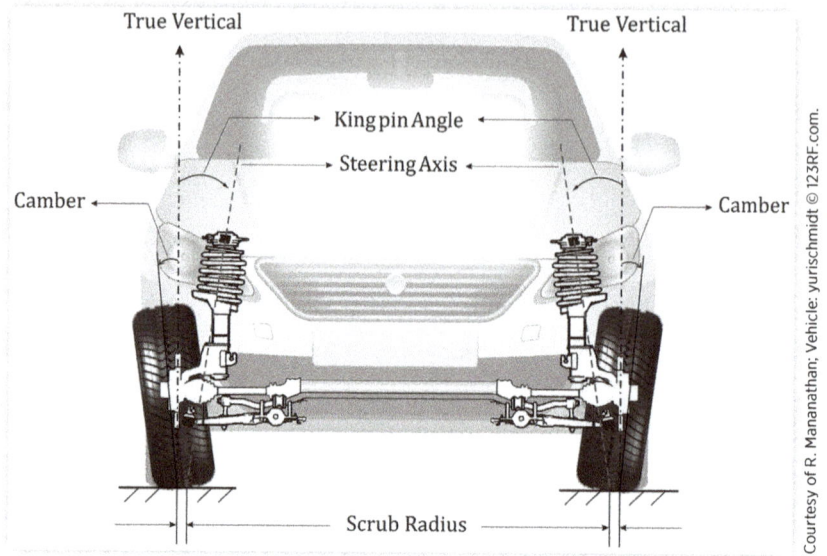

Caster

FIGURE 3.11 Positive Caster.

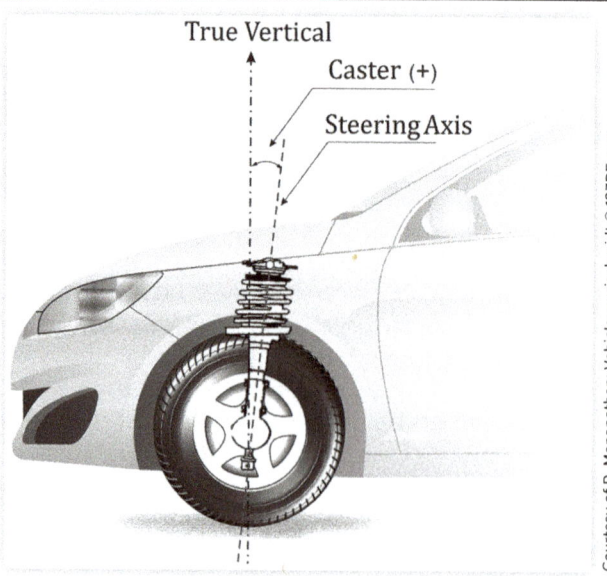

FIGURE 3.12 Negative Caster.

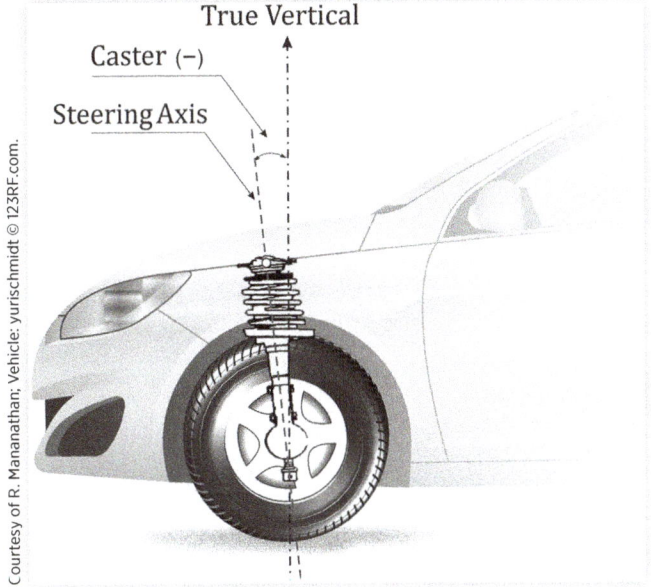

FIGURE 3.13 Caster illustration.

Caster is the angle between the true vertical and the steering axis when viewed from the side of the vehicle. As shown in Figures 3.11 and 3.12, if the inclination is towards the back of the vehicle, it is called **positive (+) Caster**, and if the inclination is towards the front of the vehicle, it is called **negative (−) Caster**. The Caster angle helps in guiding the vehicle to travel in the desired direction. This is a direction control angle.

For example, let us take a bicycle on the road, as shown in Figure 3.13.

The extension of the pipe connecting the **handlebar** and the wheel center touches the ground, little ahead of the point at which the wheel is touching the ground. This aspect guides the direction of the wheel when the handlebar is turned or makes the bicycle travel in the direction which we desire.

The same principle is applied for Caster angle in cars and other four wheelers. Besides directional control, the Caster is also responsible for the steering wheel to come back to the straight position after turning the steering wheel in curves.

Generally, the Caster will be a fixed angle in vehicles and cannot be changed. But in some vehicles, the Caster angle can also be changed. If the Caster is not proper in a vehicle, then **hard steering** will be felt. Caster will not change due to the normal wear and tear of the vehicle parts. Only when the vehicle meets with an accident, the Caster is likely to get affected.

Included Angle

Included Angle is the sum of Camber and Kingpin angles as shown in Figure 3.14.

Included angle = Camber angle + Kingpin angle

FIGURE 3.14 Included angle.

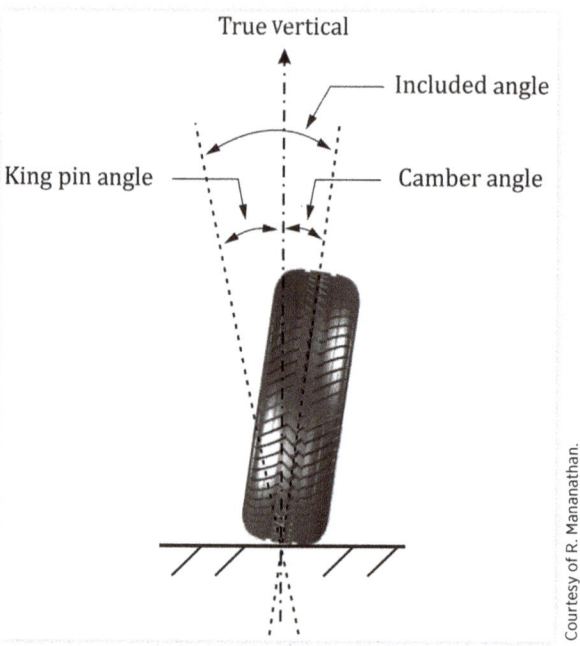

This angle is measured and given as a parameter during Wheel Alignment. The left-side Included angle and the right-side Included angle must be equal. If not, it must be corrected in the workshop.

Toe Out on Turns

When a vehicle is driven on curves, both the left and right wheels turn on a circular path. If these circular paths are completed, there will be two circles, and both the circles will have the same center point as shown in the figure. Also this point will lie on the extended line of the rear wheel center points (rear axis) as shown in **Figure 3.15**. Only when this point lies on the rear axis extension, turning on the curves will be easy and smooth. As per this principle, when a vehicle turns on curves, the inner wheel describes a smaller circle and the outer wheel describes a bigger circle. In other words, the **turning radius** of the inner wheel will be smaller than the **turning radius** of the outer wheel. The difference between the **turning angle** of the inner wheel and the outer wheel is called Toe Out on Turns (TOOT). TOOT can be measured both on the left side and on the right side of the vehicle.

FIGURE 3.15 TOOT.

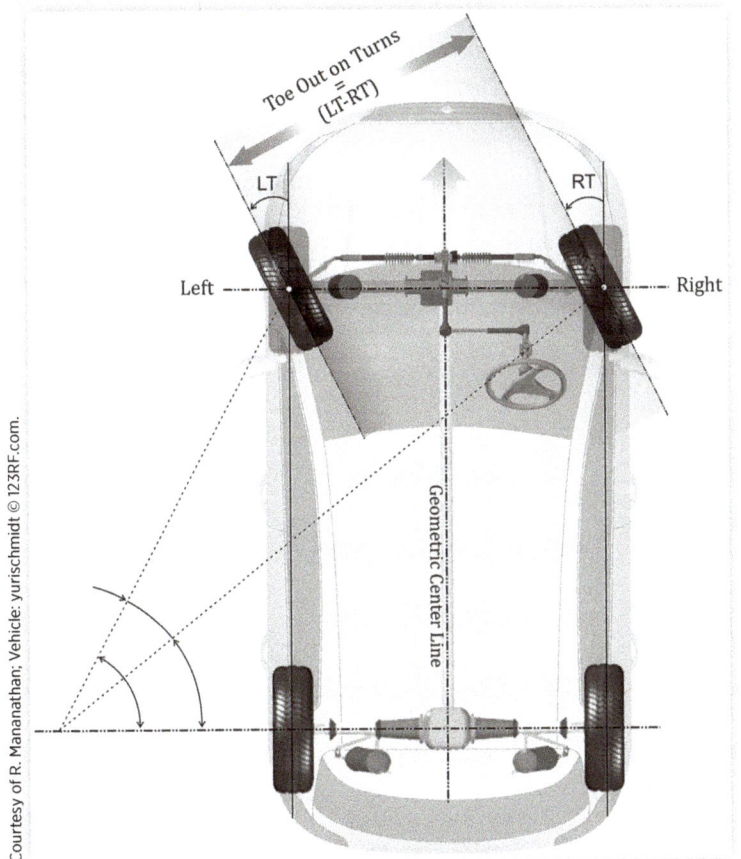

To find out, say, the Left TOOT, the following procedure is followed.
Turn the left wheel by 20° by turning the steering wheel towards the left (LT). At this point, note the angle by which the right wheel has turned (RT). Calculate the Left TOOT.

$$\text{Left TOOT} = \text{LT} - \text{RT}$$

Similarly, the Right TOOT can be found out by turning the steering wheel to the right side by 20°.

Right TOOT = RT – LT

This principle was found by Mr. Georg Lankensperger, a German, in the year 1817, and subsequently patented by Mr. Rudolph Ackermann in the year 1818. Therefore, this principle is called the **Ackermann Principle**. TOOT can be measured during Wheel Alignment, and if any correction is required, it can be done in the workshop only. Unless a vehicle meets with an accident, TOOT will not change.

Lock Angle

If the steering wheel of a vehicle is turned to one side fully, it will stop at a particular point. Beyond this point, it is not possible to turn the steering wheel. The angle between the GCL and the wheel plane at this point is called **Lock angle**. The Lock angle can be measured on both the left and right sides. The Lock angle must be equal on the left and right sides. If not equal, there is a possibility of the steering wheel getting stuck and unable to turn to the required amount when the vehicle is travelling in **hairpin bends**. This may result in accidents.

The Lock angle is likely to get changed whenever a vehicle has undergone major suspension repair in a workshop. Otherwise, this parameter is not normally measured. This can also be corrected only in workshops and not in the alignment centers.

Ride Height

FIGURE 3.16 Ride Height.

The distance between the lowermost part of a vehicle and the road is called the Ride Height, as shown in Figure 3.16. If the Ride Height is less when the vehicle crosses short radius speed breakers or sharp ups and downs on a road, the vehicle's lowest part may touch the road causing damage. If the Ride Height is more, it causes the center of gravity of the vehicle to go up, and due to this, the vehicle may topple easily. Due to this reason, a vehicle with too much luggage on the rooftop topples easily.

Vehicle designers evaluate these two aspects and determine the correct Ride Height for the vehicle. Sometimes weak suspension springs also reduce Ride Height. The Ride Height can be altered by adjusting the Upper and Lower Control arms of the suspension, and this can be done only in a workshop and not in the alignment centers, though some Wheel Aligners measure and give this parameter.

Wheel Base and Track Width

So far, we were discussing only the angles and parameters that are related to Wheel Alignment. Apart from these, we must also know about Track Width and Wheel Base.

FIGURE 3.17 Wheel Base and Track Width.

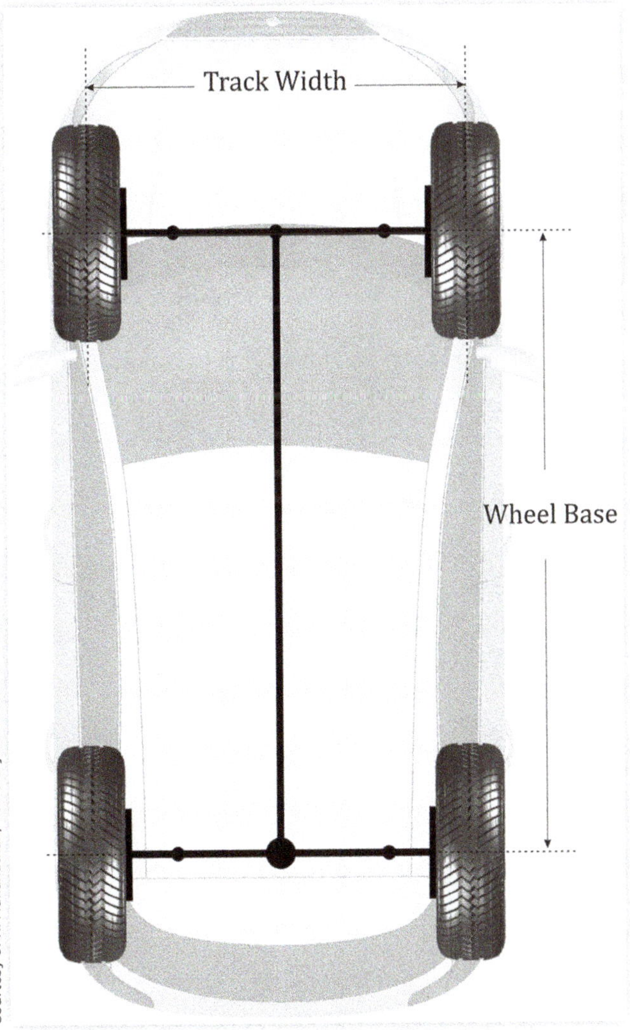

Track Width and Wheel Base represent the overall dimensions of a vehicle and are shown in Figure 3.17. These parameters will not affect the Wheel Alignment. For a three-dimensional (3D) Wheel Alignment designer, these parameters are important. These parameters help in determining the specifications of the camera and lens to be used in a 3D Wheel Aligner.

Parameters affecting Wheel Alignment

Out of the parameters described above, the Camber and Toe angles are considered to be important angles for Wheel Alignment. Though these two angles can be directly measured and the values can be taken, there are certain other parameters that interfere and affect the measurement of these two angles. It is very important to know these parameters and account for them suitably to get the correct Camber and Toe angles. The parameters that affect the measurement of Camber and Toe values are given below:

- Thrust angle.
- Wheel run-out.
- Wheel setback.

Out of the above, the Thrust angle and the Wheel Run-Out interfere heavily and affect the measurement of the Camber and Toe angles. For this reason, we can call these two as **enemies of Wheel Alignment**. Therefore, during Wheel Alignment, we must completely eliminate these enemies. If not possible to eliminate, we must compensate them suitably so that the bad effects are minimized. Before learning these, we must know an important parameter, viz., **Geometric Centerline**.

Geometric Center Line (GCL)—Definition

GCL is the line drawn perpendicular to the front axle through the middle point of the axle, as shown in **Figure 3.18**.

FIGURE 3.18 GCL.

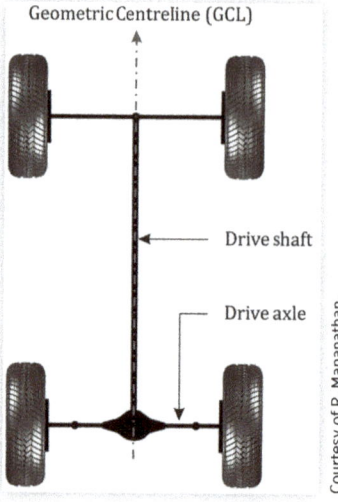

The middle point of the rear axle must lie on the GCL. Normally, the Toe angles are measured with respect to GCL only.

Thrust Angle

FIGURE 3.19 Thrust angle.

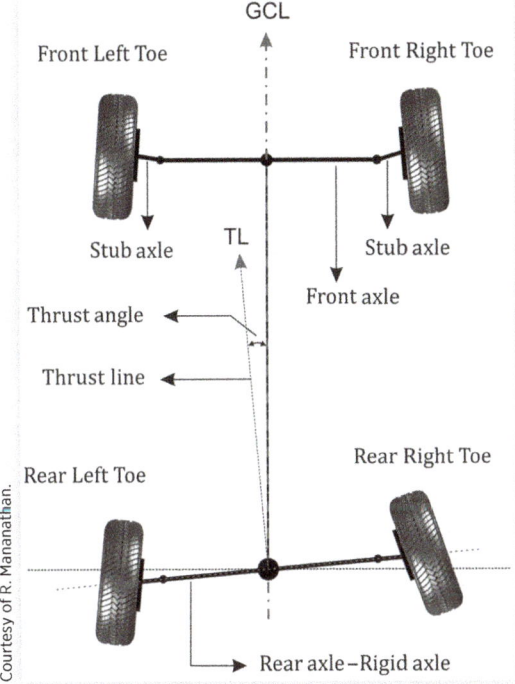

Observe **Figure 3.19** carefully. The rear axle, which is the drive axle, has the left and right wheels mounted onto it. The angle measured between the **left wheel plane** and the GCL is called Left Toe (Rear LTO). Similarly, the angle measured between the **right wheel plane** and the GCL is called Right Toe (Rear RTO).

Normally, the Rear LTO and Rear RTO must be equal in value. If equal, the bisector of the Rear Total Toe will coincide with the GCL. Under this condition, the vehicle will travel along the GCL, which is normal.

But, if these two angles are not equal, then the bisector of the Rear Total Toe will be inclined to the GCL as shown in **Figure 3.19**. The inclined line is called the Thrust Line, and the angle between the Thrust Line and GCL is called Thrust angle. If a vehicle travels under this condition, it will travel along the Thrust Line and not along the GCL. This is because the drive axle is the axle pushing the entire vehicle, and it pushes the vehicle along the Thrust line.

Now let us look at the front wheels. The Toe of the front wheels are normally set with respect to the GCL, and the front wheels will be travelling along the GCL.

Now a contradiction is created. The front wheels will be travelling along the GCL and the rear wheels will be travelling along the Thrust Line. This contradiction makes the front wheels to be subjected to abnormal friction and leads to severe tire wear.

This contradiction must be avoided to ensure the smooth running of the vehicle. The solution for this will be to make the thrust angle zero by adjusting and making the Rear LTO and Rear RTO equal in value, as per specification.

But, in reality, the rear wheels of most of the vehicles are rigidly fixed to the axle and the LTO and RTO cannot be corrected.

Even in those designs, where correction is possible, it may not be possible to make both LTO and RTO accurately equal. This will leave a **Residual Thrust Angle**, which again will lead to minimum tire wear. Under such situations what to do?

The only way out is to align the front wheel Toe angles with respect to the **Thrust Line** instead of the GCL. This will make all the four wheels run along the same line, i.e., the Thrust Line, and hence, there will not be any friction or scrubbing. To achieve this, the thrust angle measured in the rear wheels must be compensated geometrically in the front wheels.

This method of considering the Thrust angle while setting the front wheels' Toe is called "Thrust angle compensation."

This compensation is taken care of by the Wheel Alignment computers, and the technician has to simply follow the operating procedure given in the user manual of the Wheel Alignment computer manufacturer. The computer internally compensates the Thrust angle, if any, in the vehicle.

From the above explanations, it is understood that the Thrust angle compensation is essential in all kinds of vehicles to get proper Wheel Alignment. In other words, in any Wheel Alignment computer, the software must have the "Thrust angle compensation." If not, in such Wheel Aligners, even if the front Toe values are set as per specifications, there will be abnormal tire wear.

Wheel Run-Out

FIGURE 3.20 Oscillating wheel.

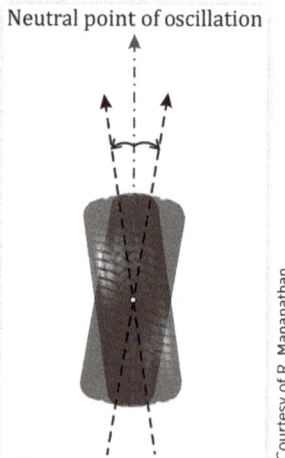

Wheel Run-Out is defined as the oscillation of a wheel with respect to its neutral plane as shown in Figure 3.20. It prevails in wheels while running on the road. Such oscillation obviously lead to tire wear. The reasons for the oscillation and the ways to manage the bad effects are given below.

Reasons for Run-out in Wheels

- Rim distortion during manufacturing.

    A wheel has two parts. A tire and a wheel rim. The tire is mounted on the wheel rim to make a wheel. The wheel rim has two circular edges. When mounted onto the axle, the inside edge of the wheel rim towards the GCL is called the **inner wheel rim** and the outside edge is called the **outer wheel rim**.

    The inner wheel rim and the outer wheel rim must be a perfect plane individually and also parallel to each other.

    But, in practice, both the rim planes may not be perfect planes and also may not be parallel to each other because of the **distortions** of wheel rims, which may happen during the production process of the wheel rims. Therefore, the middle plane of the wheel rim (neutral plane) may not be perpendicular to its own axis. If such a wheel is fitted to the axle of the vehicle, the neutral plane will not be at 90° to the axle and will be inclined to the axis (centerline) of the axle. This inclination causes the wheel to oscillate with respect to its middle position which is called as **Run-out** of wheels.

- Rim distortion happening during running.

    When a vehicle runs on the road having potholes and ridges, the wheel rim is subjected to severe forces and gets distorted over a period. This also creates **Run-out** in the wheels.

- Excess wheel bearing clearance.

    When a vehicle runs for more kilometers, the wheel bearing clearance increases due to wear and tear and results in wheel **Run-out**. This means old vehicles will have more Run-out.

- Heavy vehicles.

    Since the diameter of the wheels is generally large in Heavy Vehicles, the possible distortion during manufacturing will be more, and hence, the wheel **Run-out** will be more.

For the above reasons, even in a new vehicle, the **Run-out** will be there, and in old vehicles, **Run-out** will be more.

To understand **wheel Run-out**, look at Figure 3.21 in which a wheel is fixed to the end of a shaft and mounted on two bearings. A dial gauge is fixed to touch the outer wheel rim periphery. If the wheel is slowly rotated by hand and if there is no Run-out in the wheel, the dial gauge reading will remain static. If there is a Run-out in the wheel, the dial gauge reading will change continuously indicating a **maximum** and **minimum** value in one rotation. The difference between the maximum and minimum value in one rotation is called Run-out.

FIGURE 3.21 Run-out measurement.

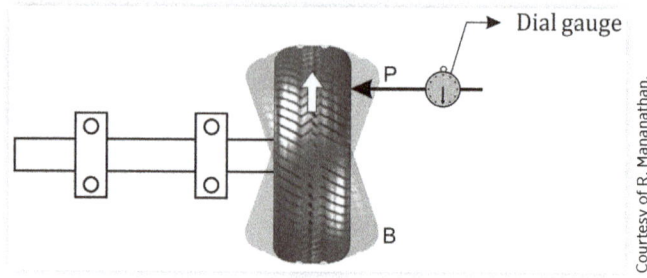

$$\text{Run-out} = \begin{Bmatrix} \text{Maximum} \\ \text{dial gauge reading} \end{Bmatrix} - \begin{Bmatrix} \text{Minimum} \\ \text{dial gauge reading} \end{Bmatrix}$$

This **Run-out** happens both in the Toe and Camber values. When the wheels run on the road, the Toe and Camber values oscillate from a minimum value to a maximum value during every rotation. This causes abnormal friction between the tire and the road resulting in excessive tire wear. Though we cannot eliminate the **Run-out** totally during Wheel Alignment, we can compensate the Run-out to reduce the tire wear considerably. Look at Figures 3.22 and 3.23.

FIGURE 3.22 Camber oscillation.

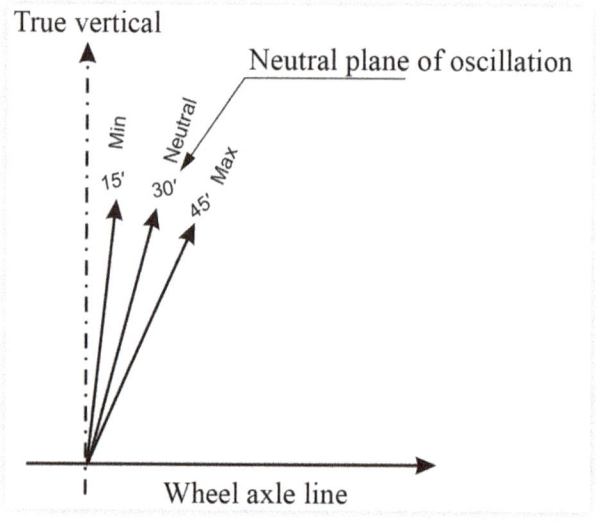

FIGURE 3.23 Toe oscillation.

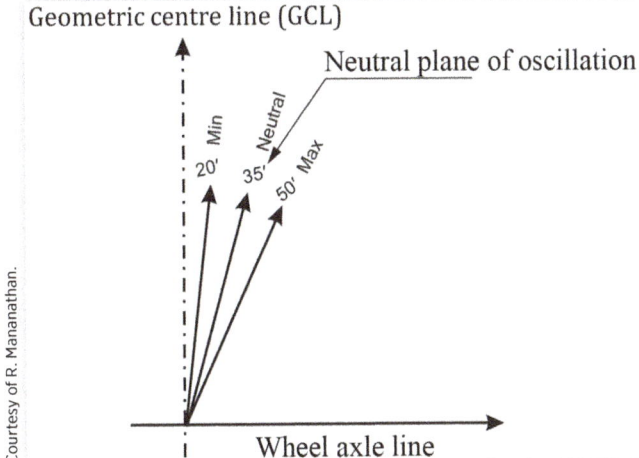

Note: The values shown in the figures are indicative only. In reality, the values will differ depending on the wheel condition.

In the above figures, the middle plane of the wheel is shown as a line. Because of Run-out, this line oscillates from a minimum angle to a maximum angle. This oscillation happens in Toe and Camber. In Toe, the oscillation is with respect to the GCL, and in Camber, the oscillation is with respect to the **True Vertical**. This oscillation happens during every rotation of the wheel. We have earlier seen that, if a vehicle runs at 120 km/hour, the wheels rotate 1820 times per minute, or, in other words, the wheel oscillates 1820 times in a minute or approximately 30 times in a second from minimum to maximum. Because of this oscillation, the tire is subjected to abnormal friction on the road and gets worn out rapidly.

When the vehicle is running on the road, this oscillation makes the Toe and Camber values a varying factor and not a static one.

Under this condition, a question arises about the correct value of the Toe or Camber of a particular wheel. For example, as shown in **Figure 3.23**, if the Toe angle of a wheel varies from 20′ to 50′, which value has to be taken as the correct Toe of the wheel? A similar question arises for Camber as well.

Since the wheel oscillates equally from its central position, this plane, called the "Neutral Plane of oscillation," has to be considered as "wheel plane."

In the case of Toe, the angle between the GCL and the Neutral Plane will be taken as Toe.

In the case of Camber, the angle between the True Vertical and Neutral Plane will be taken as Camber.

If this neutral plane is taken as the wheel plane and alignment is carried out with respect to this plane, then the wheel oscillates to an amount of **oscillation/2** from its neutral position in both directions. Instead, if the alignment is performed taking the minimum position or the maximum position of oscillation as wheel plane, the Toe and Camber values in every rotation will exceed the set specification by more value and will result in excessive tire wear. Therefore, taking the **Neutral Plane** as the wheel plane is the best way for Toe and Camber alignment.

But, when a vehicle comes for Wheel Alignment and gets parked in the bay, its four wheels may not be in the neutral position. They may stand in any position between the minimum and maximum positions of the oscillation. To get the correct Toe or Camber, the four wheels of the vehicle must be parked in the neutral position. This is practically impossible because the Neutral Plane is not known.

The Run-out program in a Wheel Alignment computer calculates and finds out the Neutral Plane of every wheel, unmindful of its parking condition. This is done by an intelligent software.

There are many methods of finding the **Neutral Plane** in the Run-out program of a Wheel Alignment computer, which will be discussed in detail later in this chapter. Whatever may be the method, the fundamental principle of Run-out measurement and finding the Neutral Plane is the same for all the methods. Now we will understand how the Run-out program works.

Principle of Run-Out

Since the **Run-out** values oscillate between a maximum and minimum value in one rotation (360°), this oscillation will be a **sine curve**, as shown in Figure 3.24.

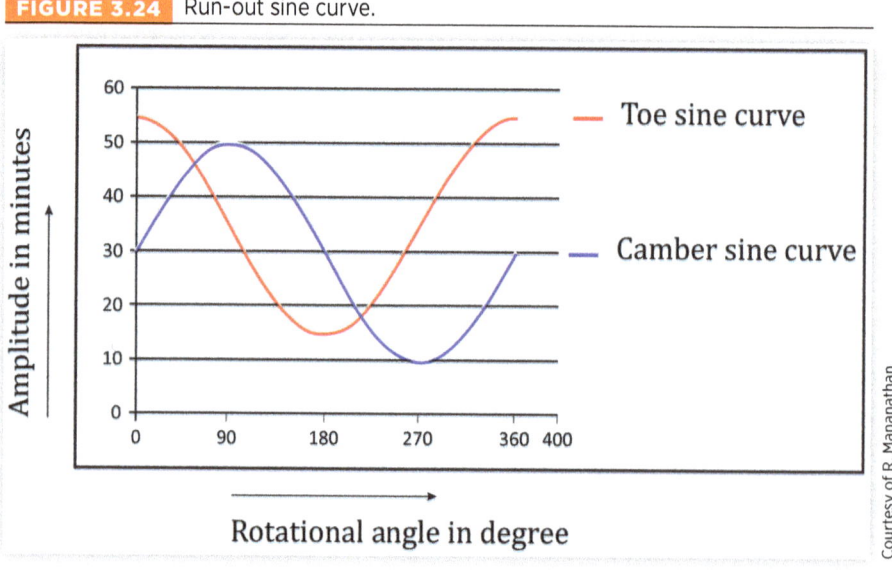

FIGURE 3.24 Run-out sine curve.

By using the formula for sine curve, it is possible to find out the neutral plane of oscillation. This formula is a complex one and requires certain data from the cameras (sensors) to draw the sine curve of a particular wheel's oscillation. This data is acquired in many ways:

1. Maxima - Minima method.
2. Four Point method.
3. 30° Push method.
4. 180° Push method.

In the first two methods, every wheel has to be lifted and rotated manually by using a jack to get the data. This requires a lot of human effort. The third and fourth methods are easy methods in which, instead of lifting the wheels, the vehicle has to be pushed by 30° or 180° of wheel rotation. During this push, the changes in the Toe and Camber readings are acquired. Using this data, the position of **Neutral Plane** with respect to the parked plane is calculated. This 30° or 180° Push method is followed in 3D Wheel Aligners.

During this Run-out measurement, the angle between the Parking Plane and Neutral plane is found out, and this will be suitably accounted for in the Camber and Toe readings. Correcting the Camber and Toe values using this value is called **Run-out Compensation**. This is very important for Wheel Alignment.

If **Run-out compensation** is not carried out properly, the following bad effects will happen:

1. Abnormal tire wear.
2. Steering Cross after alignment.
3. Side Pulling while driving.

In practice some vehicles may have very high **Run-out** in the wheels. If Wheel Alignment is carried out in such wheels having excess Run-out even with proper Run-out compensation, abnormal tire wear may still be there. Such vehicles must be sent to the workshop for necessary corrections to reduce the excess **Run-out** in wheels.

> **Note:** Both Thrust angle and Wheel Run-Out are considered to be ENEMIES of Wheel Alignment. In many vehicles these parameters cannot be corrected and eliminated during Wheel Alignment, but can be compensated. By carrying out **Thrust angle compensation** and **Run-out Compensation**, the bad effects can be greatly reduced.

A good Wheel Alignment computer takes care of the above compensations in its design itself. The technician's job is just to carry out Wheel Alignment as per the procedure given in the User Manual of the Wheel Aligner manufacturer.

Wheel Setback

The left wheel and the right wheel mounted at the end of the axle in a vehicle must be on the same axis (centerline of the axle). Due to manufacturing defects or due to accidents, the left and right wheels may not be in the same centerline of the axle.

FIGURE 3.25 Setback.

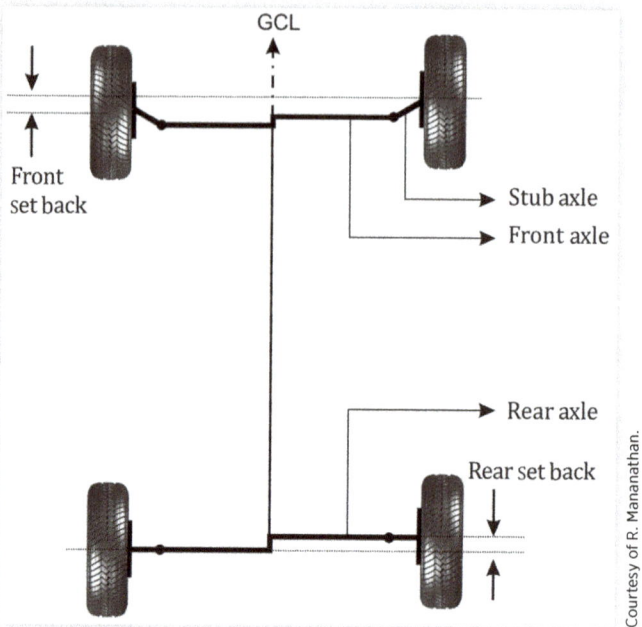

Any one wheel might be offset from the centerline, as shown in Figure 3.25. This offset is called **Wheel Setback**. This Setback is measured in millimeters. Some Wheel Alignment computers give this in angle (minutes) deviation from the axis.

The **Wheel Setback** does not affect the Wheel Alignment. This does not lead to excessive tire wear. But it causes difficulty in the measurement of Toe readings in certain Wheel Alignment computer designs. To eliminate this, the design engineers measure the Setback and compensate to get the correct Toe readings. But most of the Wheel Alignment computers measure Setback and give this parameter as output. If the Setback is more, it can be corrected in the workshop only.

Wheel Alignment angles measuring technology and Its advancement

The following angles are measured during Wheel Alignment (Table 3.1).

TABLE 3.1 Wheel Alignment angles.

1. Camber	Wheel angles
2. Toe	
3. Caster	Steering axis angles
4. Kingpin angle	

The other Wheel Alignment parameters are derived from the above data.

Measurement of these angles involves high-tech electronic sensors, circuits, and software applications. Many advancements have taken place in the last few decades in the technology of measuring the **Toe** parameter. Still researches are going on to achieve the following:

- To improve the accuracy of readings.
- To reduce the time taken for the alignment.
- To make the job simple for mechanics.
- To reduce human effort.
- To accommodate new design of vehicles.

Tables 3.2 and 3.3 give the technologies used in the design of Wheel Alignment computers.

TABLE 3.2 Camber—Technology development.

Before 1990	Potentiometer with pendulum
1990-2000	Liquid potentiometer
2000-2010	Onboard MEMS sensor
2010-2018	3D image processing
After 2018	Touchless technology (still not in full use)

TABLE 3.3 Toe—Technology development.

Before 1990	1. Potentiometer with thread 2. Laser technology
1990-2000	Optical technology
2000-2010	CCD camera technology
2010-2018	3D image processing technology
After 2018	Touchless technology (still not in full use)

Caster and Kingpin

These two parameters are measured using the same sensors deployed for the Camber. Therefore, the technology and period given for Camber can be taken for Caster and Kingpin.

Charge Coupled Device (CCD) Technology

Though 3D technology has come up widely for use, CCD technology is also used by some users for cost considerations. Therefore, let us understand briefly about the CCD technology.

In 3D technology, a single image plate alone is capable of measuring all three angles in the three directions. But, in CCD technology, two separate sensors are used to measure the three angles as given below:

- CCD cameras are used to measure the Toe angles.
- MEMS sensors are employed to measure the Camber, Caster, and Kingpin angles.
- To process the output of these sensors, a microcontroller board or a minicomputer is deployed.
- The output of this minicomputer is displayed in the computer display.
- **CCD Wheel Aligner.**

FIGURE 3.26 CCD wheel aligner.

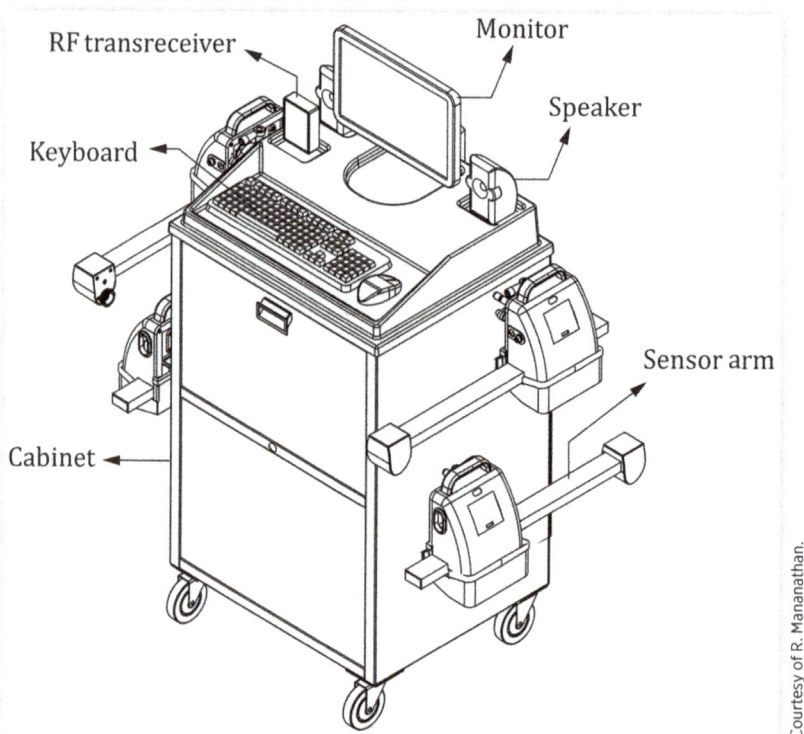

There is one sensor arm for each wheel, and in total, there are four sensor arms for all four wheels. These sensor arms are mounted onto the wheels through the wheel brackets like the image plates are mounted in 3D technology.

The CCD Wheel Aligner shown in Figure 3.26 is an equipment consisting of the following parts:

- **CCD Sensor Arm.**

An assembly that contains all these sensors and electronic boards is called a sensor arm and is shown in **Figure 3.27**:

FIGURE 3.27 CCD sensor arm

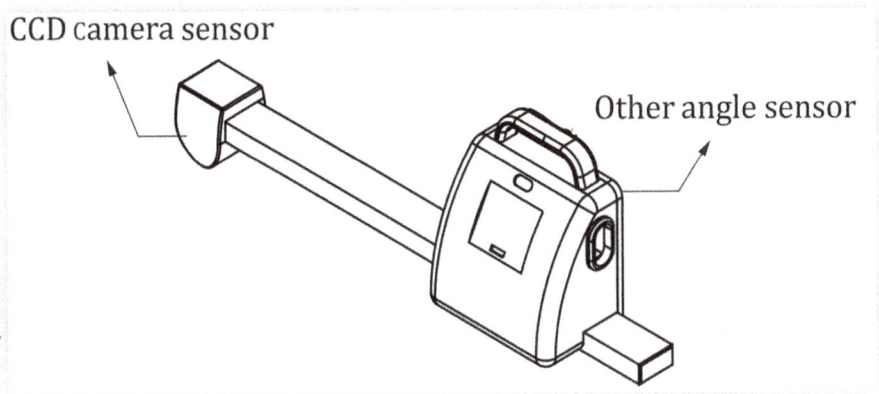

Courtesy of R. Mananathan.

- Microcontrollers and computer.
- Display unit.
- Power supply unit.
- Accessories like wheel clamps, rotary plates, etc.
- A cabinet to accommodate the above.

These CCD Wheel Alignment computers are capable of measuring all Wheel Alignment parameters. In the procedure of carrying out Wheel Alignment, there is no difference between a CCD Wheel Aligner and a 3D Wheel Aligner. Still, for reasons described below, the 3D Wheel Aligners are considered to be far better than CCD Wheel Aligners:

- In CCD technology, at the rate of three sensors per sensor arm, there are twelve sensors deployed for measuring all Wheel Alignment parameters. Also equivalent electronic parts are there to support the design.

  In 3D design, a single image plate measures all three angles at a time, and only four such image plates are used to measure all the wheel alignment parameters.

- In CCD technology, rechargeable batteries are necessary for each sensor arm to make the above electronic parts function. Otherwise, an electric cable must be connected to each sensor arm to make the arms function, which will be very cumbersome to operate.

  In 3D design, battery or electric connection is not required for the image plates.

- Since a lot of electronic components are there in a CCD sensor arm, which is supposed to be removed and refixed for every alignment, the sensor arms are likely to get frequent failures.

Since there are no electronics in 3D design, the failure possibilities are very less.

For the above mentioned reasons, 3D Wheel Aligners are preferred compared to CCD Wheel Aligners. In the future the CCD usage will vanish from the field.

Since the CCD technology will disappear from the field and since the **Touchless** technology has not yet been successful, 3D technology will prevail in the field for more time. Therefore, further explanations in this book will be based on the widely used 3D technology only.

3D Technology

3D Planes

Out of the four angles, i.e., Toe, Camber, Kingpin, and Caster, the Toe angle is measured in the Horizontal Plane, the Camber, and Kingpin angles are measured in the Vertical Plane, and the Caster angle is measured in the Transverse Plane. This means, for measuring Wheel Alignment parameters, angles have to be measured in 3D planes. Let us assume that the three planes are assigned notations as X, Y, Z, respectively:

- X—Horizontal Plane.
- Y—Vertical Plane.
- Z—Transverse Plane.

When we see a wheel by standing in front of a vehicle, we can understand the 3D planes from **Figure 3.28**.

FIGURE 3.28 3D planes.

For Wheel Alignment purposes, the angle at which a tire is inclined in the above three directions has to be measured for all four wheels. This means twelve sets of data have to be acquired from the four wheels of the vehicle.

Image Plate

To measure the angles in the three directions, an Image Plate is used in each wheel as shown in Figure 3.29. When the Image Plate is mounted to a wheel using Wheel Clamps, the Image Plate represents the angles of the wheel in three planes, i.e., X, Y, Z. In other words, if the angle of the Image Plates is measured, it is as good as measuring the angles of the wheel.

FIGURE 3.29 Image plate fixed on the wheel rim.

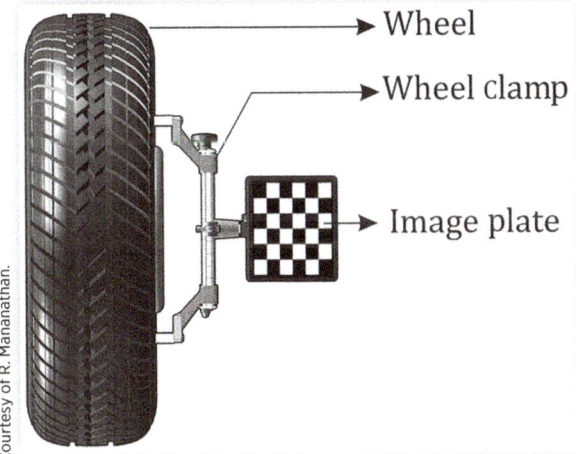

Courtesy of R. Mananathan.

In 3D technology, these Image Plates work along with a digital camera. These Image Plates have a reflective layer pasted on their surface. These reflective layers contain eighteen black squares and eighteen white squares called patterns. (Some designers use circular patterns and some other designers use different shapes as patterns based on their design considerations.) But all the patterns serve the same purpose of measuring the angles in three planes.

Camera Sensor

High-resolution digital cameras are used to capture the images of the image patterns. The cameras are mounted in front of the Image Plates at a predetermined distance from the Image Plates, in such a way that the Image Plates fall within the field of vision of the cameras. If the Image Plate is kept in front of a digital camera, exactly parallel to the camera screen, i.e., without any tilt in any direction, the image captured in the camera will be of perfect squares as seen in Figure 3.30. If the camera is tilted, say, in the transverse (Z) direction, the image in the camera will become a rectangle, as shown in Figure 3.30a.

FIGURE 3.30 Image plate parallel to the camera.

FIGURE 3.30A Image plate tilted in the Z direction.

If the Image Plate is tilted in the X, Y directions, the images will become different in shape depending on the amount of tilt in each direction. All the wheels in a vehicle will always be tilted in all three directions, and if the image plate is mounted to the wheels, then the images displayed will be having a shape proportional to the wheel angles. If the camera is connected to a computer, we can see these images on the computer screen. The amount of change from the square shape of the image indicates the amount of wheel tilt. By sensing the change in the shape of the square images, the microcontroller connected to the camera is able to calculate the exact angle of the wheel tilt in each direction, i.e., in the X, Y, Z directions. This is done using an image processing software.

Measuring the Angles in Vehicles Using 3D Technology

Figure 3.31 shows a vehicle in which the image plates are mounted on all four wheels and parked in front of the cameras. Now the vehicle is ready for alignment.

FIGURE 3.31 All image plates mounted on wheels and ready to carry out the alignment.

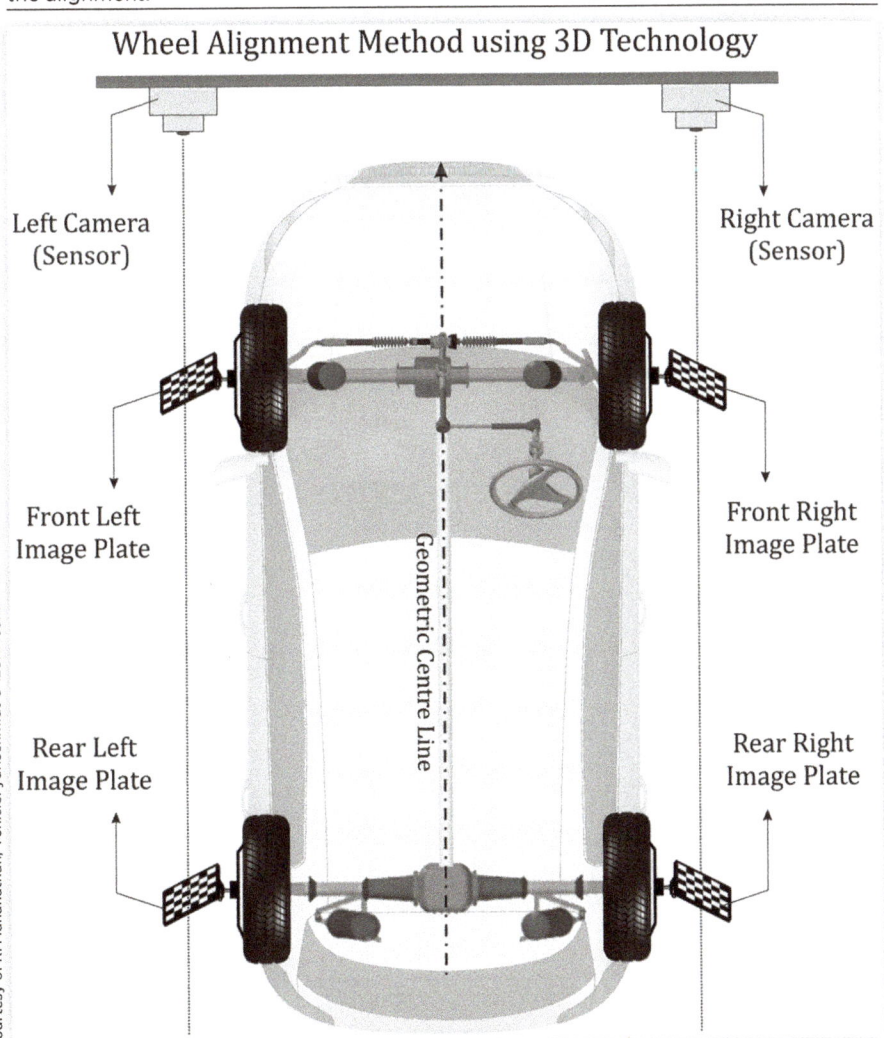

There are two digital cameras, one on the left side and the other on the right side in front of the vehicle. The left-side camera receives the images from the left-side wheels of the vehicle (Front and Back). Similarly, the right-side camera receives the images from the right-side wheels of the vehicle (Front and Back). Since the Image Plates are tilted to the extent of Camber and Toe in each wheel, the images on the camera screens will not be in a square shape, but they will be elongated rectangles in shape. By capturing these images and their changes in shape, the X, Y, Z angles of each wheel are calculated by the camera and sent to the computer for further processing.

Now the computer will have three values of angles for each wheel, totaling twelve values. On getting these twelve values, the computer processes them using a Wheel Alignment software consisting of many mathematical formulae and gives an output of all Wheel Alignment parameters for each wheel individually. These output angles will be **Run-out compensated** and **Thrust angle compensated**.

A Wheel Alignment computer measures all these parameters, and such a typical Wheel Alignment computer is described in the next chapter.

3D Wheel Alignment Computer and Its Accessories

A 3D Wheel Alignment computer is shown in Figure 3.32 with its main parts. Today many kinds of Wheel Alignment computers are available in the market, at prices ranging from US$ 3000/- to US $20000/-. These aligners calculate the Wheel Alignment parameters accurately and display them in the system. Also they give a **printout** showing the specifications, the initial and final readings of alignment of a particular vehicle.

FIGURE 3.32 3D Wheel Alignment computer.

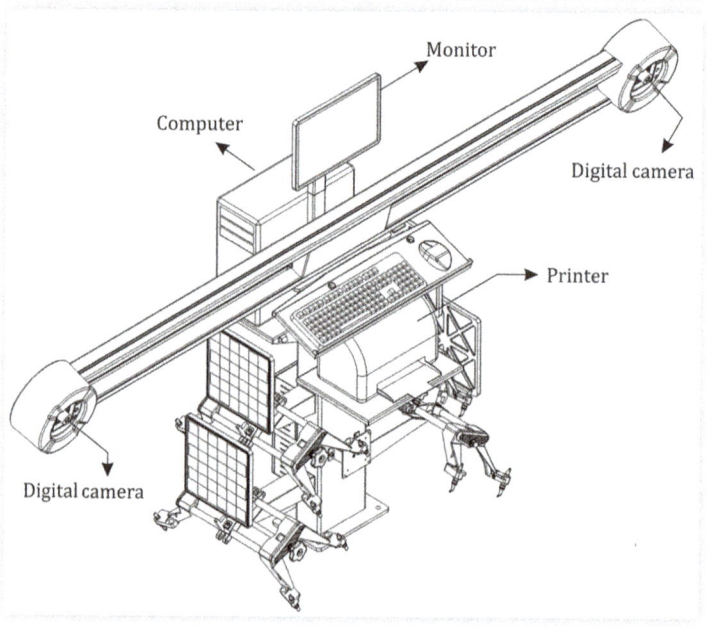

3D Wheel Alignment Computer

A Wheel Alignment computer uses various technologies for measuring the different angles of Wheel Alignment:

Technologies used

- Sensors/Cameras.
- Lens and optical technology.
- Infrared technology.
- Microcontrollers.
- Hardware.
- Embedded software.
- Front-end software.
- Animations.
- Image processing software.
- Mathematical algorithms.
- Automobile technology.
- Mechanical designs.

Though so many technologies are used, eventually this aligner must be used by the technicians who carry out Wheel Alignment. Unless these technicians understand and use the aligners properly, the possibility of getting a wrong result is high. This is true even when using a high-cost Wheel Aligner.

Keeping the above point in mind, this book is written to help the technicians and engineers to understand what is Wheel Alignment and Wheel Balancing and how to carry out these activities knowledgeably.

Besides the parts shown in **Figure 3.32**, the following accessories are also used in a Wheel Alignment computer.

Since the functions of cameras and image plates have been explained already, the other accessories used in Wheel Alignment will be explained here (**Table 3.4**).

TABLE 3.4 Accessories used.

Description	Nos
Rotary plates	2
Wheel clamps	4
Steering lock	1
Brake pedal lock	1
Rear wheel slider	2
Wheel stopper	2

Rotary Plate (Turn Table)

While carrying out Wheel Alignment, the Rotary Plate as shown in **Figure 3.33** will be kept below the front left wheel and front right wheel. When Caster, Kingpin, TOOT, and Lock Angles are measured, the front wheels have to be turned towards the left and right directions using the steering wheel. To allow the wheels to turn easily without any friction, these rotary plates are designed with a set of small-sized balls underneath the top plate. These balls may be made of steel or special plastic material to withstand the load. A spring assembly inside the rotary plate ensures the top plate returns to its middle position automatically when the load is removed.

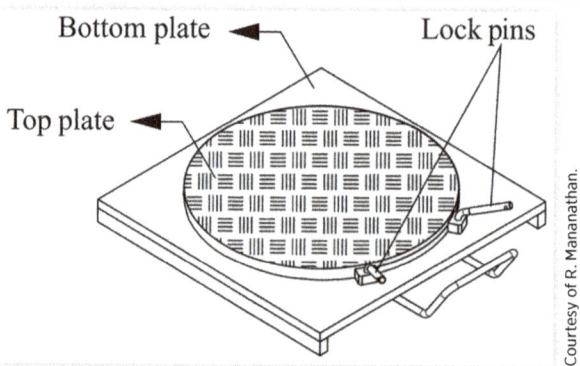

FIGURE 3.33 Rotary plate.

Since these Rotary Plates are kept below the wheels, dirt is likely to get deposited in between the plates affecting the free movement. At least once in fifteen days, the Rotary Plates must be dismantled and cleaned to remove the dirt deposits, if any. The springs and balls must be inspected for any damage. If the balls are made of plastic, they must be inspected for any damage or breakage, and if the balls are made of steel, the bottom plates must be inspected for any damage or impression by the steel balls.

> Caution: The following problems will be faced in Wheel Alignment if the above precautions are not followed.

- Run-out value will get affected resulting in a wrong Toe and Camber.
- Caster and Kingpin angle readings will be wrong.

In short, all the Wheel Alignment angles will be affected if the Rotary Plate is not maintained properly. Therefore, it is very important to service and maintain the rotary plates periodically.

Wheel Clamp (Wheel Bracket)

Wheel Clamp as shown in Figure 3.34 is also an important accessory for Wheel Alignment. During Wheel Alignment, one wheel clamp will be fixed on each wheel. In each wheel clamp, an image plate is fixed which will reflect the wheel angle.

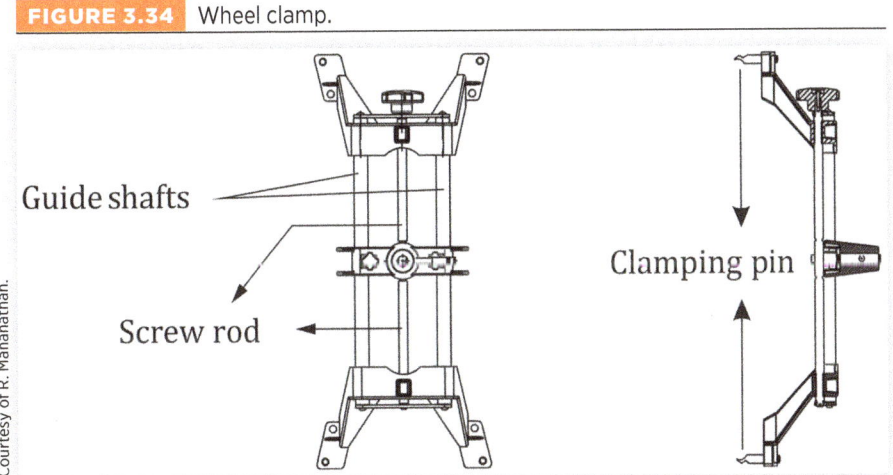

FIGURE 3.34 Wheel clamp.

For this reason, the wheel clamp and image plates must be mounted onto the wheels very carefully observing the following precautions:

- The clamping pins in the wheel clamp must be mounted to the wheel rim where there is no dent or distortion (choose the correct location).
- All four clamping pins must be pressed uniformly for seating on the wheel rim without any gap.
- The image plate must be mounted onto the wheel clamp ensuring horizontality through a spirit level available on the image plate.
- While fixing the wheel clamp, care must be taken not to drop the wheel clamp or the image plate on the ground. If the wheel clamp is dropped, the image plate may get damaged and the consequences will be improper alignment readings.
- The guide shaft and screw rod in the wheel clamp should not have any bend or distortion.

Fixing the wheel clamp on the wheel rim is a human activity and also a critical activity for Wheel Alignment. This activity depends on the training of the technicians carrying out alignment. If the wheel clamp is not mounted properly onto the wheel rim, the following effects will be there on the Wheel Alignment results:

- Toe and Camber readings will get affected (since the effect is severe, abnormal tire wear will be there).
- Steering Cross may be there after alignment.
- Side pulling may be there while driving.

Therefore, the Wheel Clamps must be cleaned and serviced every fifteen days. Since the fixing of the wheel clamp has to be done in every vehicle, the technicians must be trained well on fixing the wheel clamps to the different wheel rims.

Steering Wheel Lock

This accessory shown in **Figure 3.35** is used for locking and keeping the steering wheel in a fixed position. Particularly, while carrying out **Run-out** operation, the steering wheel should not move or turn even a little bit. The Steering Lock helps in firmly holding the steering wheel in one position. It must be fitted on the steering wheel before **Run-out** operation. If the steering wheel got disturbed or moved during **Run-out** operation, wrong **Run-out** readings will be calculated and all the Wheel Alignment parameters will get greatly affected. While adjusting the front wheel angles, also the steering lock must be fixed. Technicians must understand this during their training.

FIGURE 3.35 Steering wheel lock.

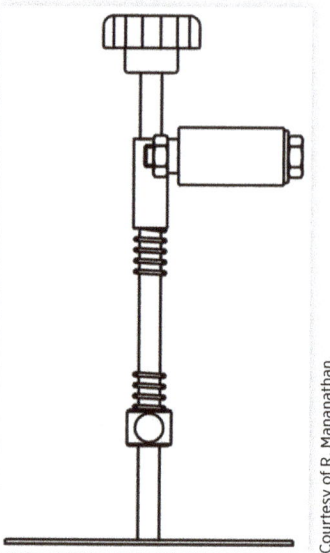

Courtesy of R. Mananathan.

Brake Pedal Lock

FIGURE 3.36 Brake pedal lock.

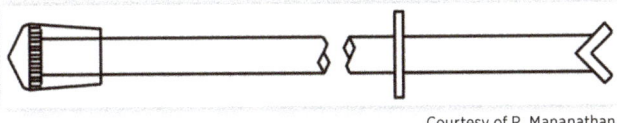

Courtesy of R. Mananathan.

This accessory shown in Figure 3.36 is used for keeping the brake pad of the vehicle under pressed condition or in brake applied condition. Particularly, when the steering wheel is turned towards the left and then towards the right for measuring the Caster and Kingpin angles, the vehicle should not move. To ensure this, the Brake Pedal Lock is fitted in such a way that the brake is strongly kept pressed before this operation. If the Brake Pedal Lock is not fitted properly, wrong Caster and Kingpin readings will be given by the aligner.

Rear Wheel Slider

FIGURE 3.37 Rear wheel slider.

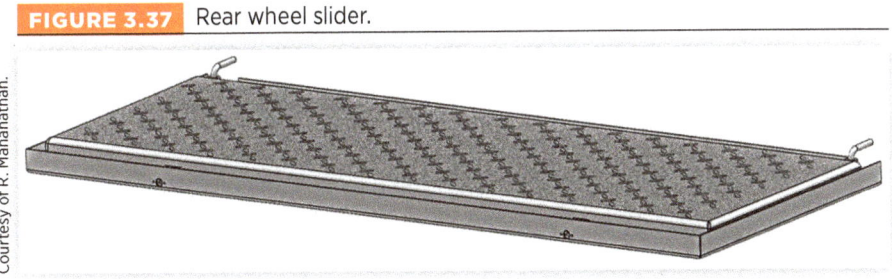

This accessory shown in Figure 3.37 is kept below the rear wheels of the vehicle during alignment to ensure the rear wheels attain proper position for Wheel Alignment. The springs available inside the rear wheel sliders help in keeping the rear wheels in the correct position.

Wheel Stopper

During Run-out measurement, the vehicle has to be pushed **back and forth** by 30°. With such pushing, the stoppers, as shown in Figures 3.38 and 3.39, help in preventing the vehicle from being pushed beyond limits. These stoppers are supplied as an accessory along with the 3D Wheel Aligners.

FIGURE 3.38 Wheel stoppers.

FIGURE 3.39 Wheel stoppers in position.

3D Wheel Aligner—Installation

While performing Wheel Alignment, it becomes necessary to carry out corrections of the angles as per specifications. For this, the technician must have access to the bottom of the vehicle for effecting the corrections easily. To facilitate this, certain infrastructure is required in the Wheel Alignment centers.

Wheel Alignment Pit

A **pit** can be constructed in which the mechanic can go down and access the bottom of the vehicle for effecting the corrections.

A pit has to be constructed, as shown in **Figure 3.40**, in the location where the vehicle will be parked for alignment. Since the mechanic has to go down and work, the pit must have sufficient depth and steps to go down. While constructing the pit, it must be ensured that all the locations where the wheels will be seating are in the same plane (level). For this, the following points have to be observed during the construction of the pit.

Automobile Wheel Alignment and Wheel Balancing

FIGURE 3.40 Wheel alignment pit.

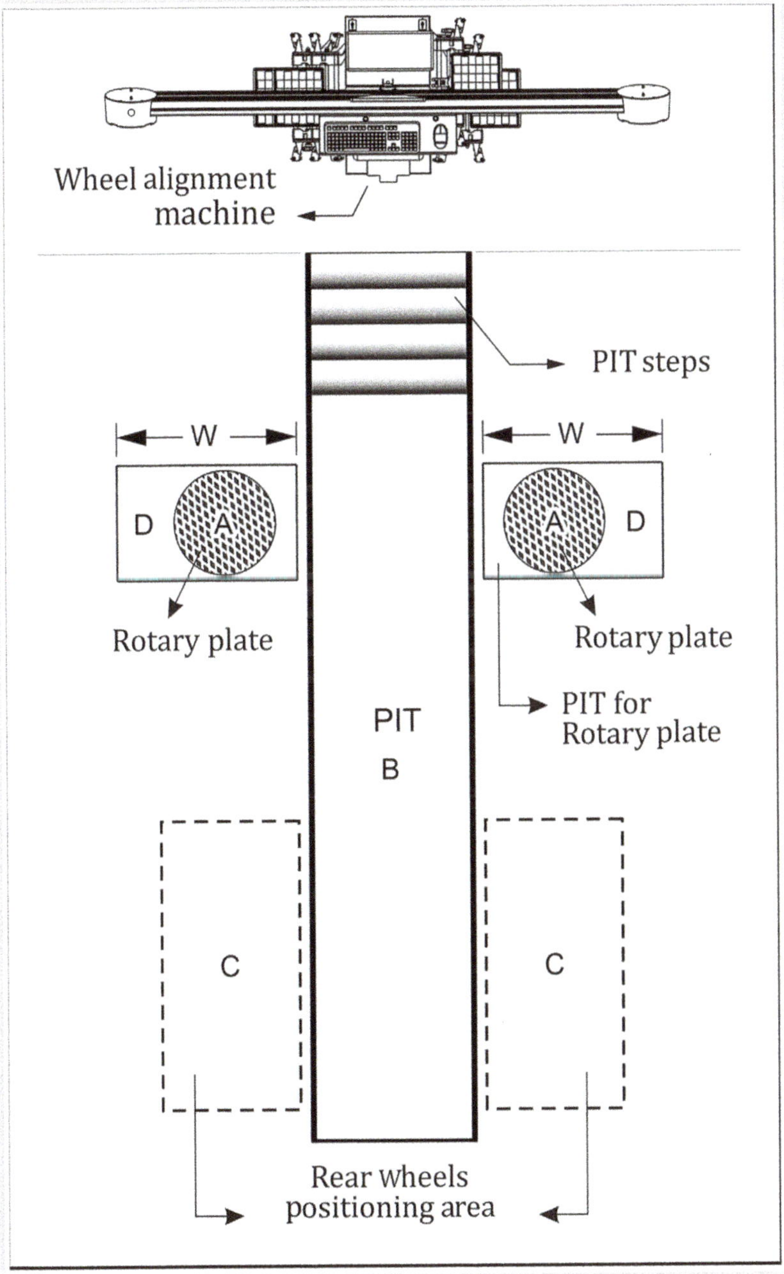

Wheel alignment machine

PIT steps

W

D A

Rotary plate

W

A D

Rotary plate

PIT for Rotary plate

PIT B

C

C

Rear wheels positioning area

Courtesy of R. Mananathan.

Note: The above drawing is a general layout for a pit. If the Wheel Aligner manufacturer has given a specific drawing for constructing the pit, it must be followed.

A—Rotary Plates: Left and right
Normally, the Rotary Plates are kept in a depression in such a way that the top surface of the Rotary Plates is on the same level as the floor.

D—Depth of depression for placing the Rotary Plates
The left and right depressions must be on the same level. It is sufficient if water level is maintained between the left and right sides.

W—Width of depression
The width must be sufficient enough to accommodate different track widths of small and big vehicles. The depression can be finished with a granite surface for easily moving and positioning the Rotary Plates under the wheels of different vehicles having different Track Widths.

C—Rear wheels seating locations
Depression to be made to accommodate the Rear Wheel Slider.

A&C: Both of these must be on the same level and can be finished using the water level.

C&D: It is advisable to lay granite in the C and D locations where the wheels get seated during alignment.

Wheel Alignment Lift

In this method, the alignment will be carried out in a lift. For this, either a Scissor Lift or a Four Post Lift can be used. In Figure 3.41, a Scissor Lift is shown as an example. While installing the lift, care must be taken to ensure the lift is placed on a leveled surface without any undulations.

FIGURE 3.41 Wheel alignment Scissor Lift.

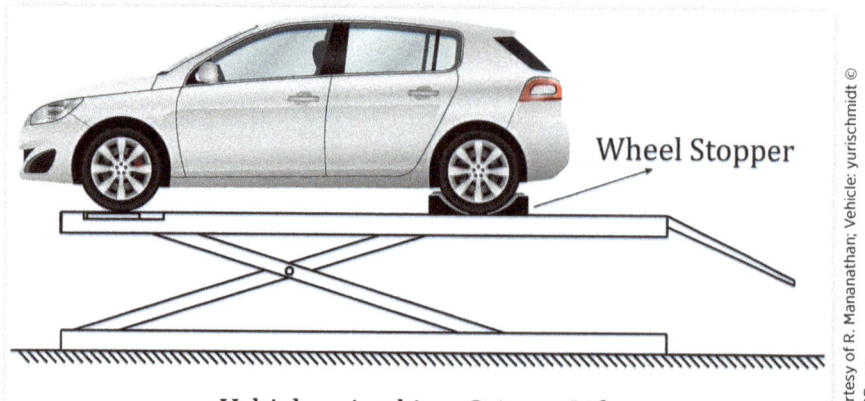

Vehicle raised in a Scissor Lift

Courtesy of R. Mananathan; Vehicle: yurischmidt © 123RF.com.

While performing Wheel Alignment, the initial condition of the wheels must be measured to start with. For this, the lift must be kept in its lowermost position and the vehicle must be parked on the lift in this lowered position. Once the initial readings are measured, the lift has to be raised to a suitable height for effecting corrections of the necessary angles.

Wheel Aligner—Installation Procedure

The Wheel Aligner must be installed in front of the pit or lift as shown in Figure 3.42. The installation has to be carried out as per the procedure given in the **Service Manual** supplied by the manufacturer of the Wheel Aligner.

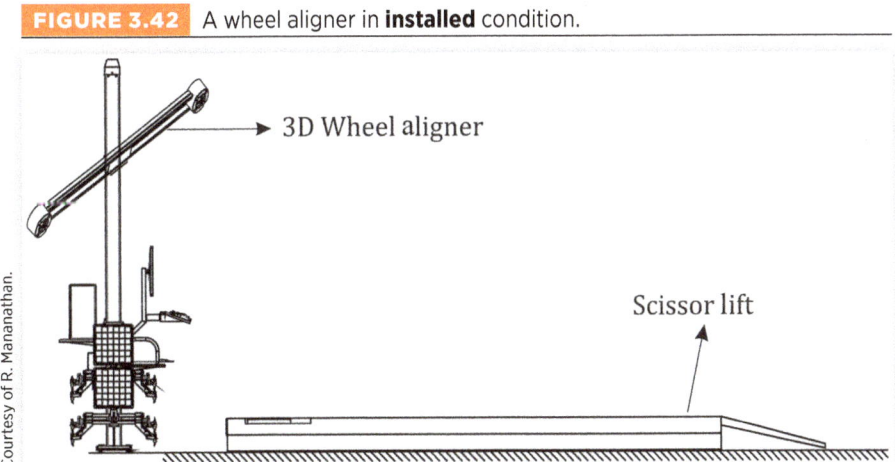

FIGURE 3.42 A wheel aligner in **installed** condition.

3D Wheel Aligner—Calibration

The first job to be done after installing a Wheel Aligner will be **Calibration**. This is a very critical and important activity. This must be carried out with lot of care.

- **What Is Calibration?**

    Let us assume that we are manufacturing Electronic Weighing Machines. During production, immediately after assembling the machine, if a 50-gram weight is placed on it, it will not display 50 grams. This is because the weighing machine has not been calibrated or, in other words, the machine has not been taught to display 50 grams as 50 grams. It may display either 45 grams or 55 grams or any other value.

    The process of making the weighing machine display the correct value of a known weight placed on it is called **Calibration**.

Once the calibration is done, the weighing machine will measure and display all the weights placed on it correctly. This calibration is done internally by a software. This is essential for all electronic gadgets including Wheel Alignment computers.

- **Factory Calibration.**

    In Wheel Aligners, the process of **calibration** will make the aligner give correct angles. Normally, the calibration will be done in the factory during manufacturing under ideal conditions. This is called **Factory Calibration**.

- **Field Calibration.**

    While installing the Wheel Aligner in the field, the same ideal environment may not be available in terms of floor level, pit level, or lift level. For this reason, **calibration** is done after installing a Wheel Aligner in its working location. This is called **Field Calibration**.

Most Important: Every manufacturer of a Wheel Aligner will prescribe a procedure for carrying out the Field Calibration of his product. It is very important to strictly follow the same procedure without any deviation.

A general procedure of how a typical 3D Wheel Aligner is calibrated after installation is given below. For Wheel Alignment, the angles have to be measured in the following three planes:

- X—Horizontal Plane.
- Y—Vertical Plane.
- Z—Transverse Plane.

In 3D technology, only one image plate is capable of measuring all three angles in the above three planes. This aspect helps in carrying out the calibration of all three angles in a single operation. This is an advantage of 3D Wheel Aligners compared to CCD Wheel Aligners.

Calibration Procedure of a 3D Wheel Aligner

Every manufacturer would have supplied a Calibration Kit (**Figure 3.43**) for carrying out Field Calibration. Normally, the Calibration Kit will be a rectangular frame assembly with its four corners manufactured precisely having 90°. The kit has provisions to accommodate four image plates. When these image plates are fitted in the four corners, their center points are also precisely at 90°, forming a perfect rectangle. As seen in the figure, the assembled kit along with the image plates will be placed in front of the Wheel Aligner, either on the pit or on the lift, as the case may be. The kit must be placed in such a way that the centerline of the kit coincides with the centerline of the pit or lift. This will be guided by the Wheel Aligner through its screen.

FIGURE 3.43 Image plates mounted on calibration jig.

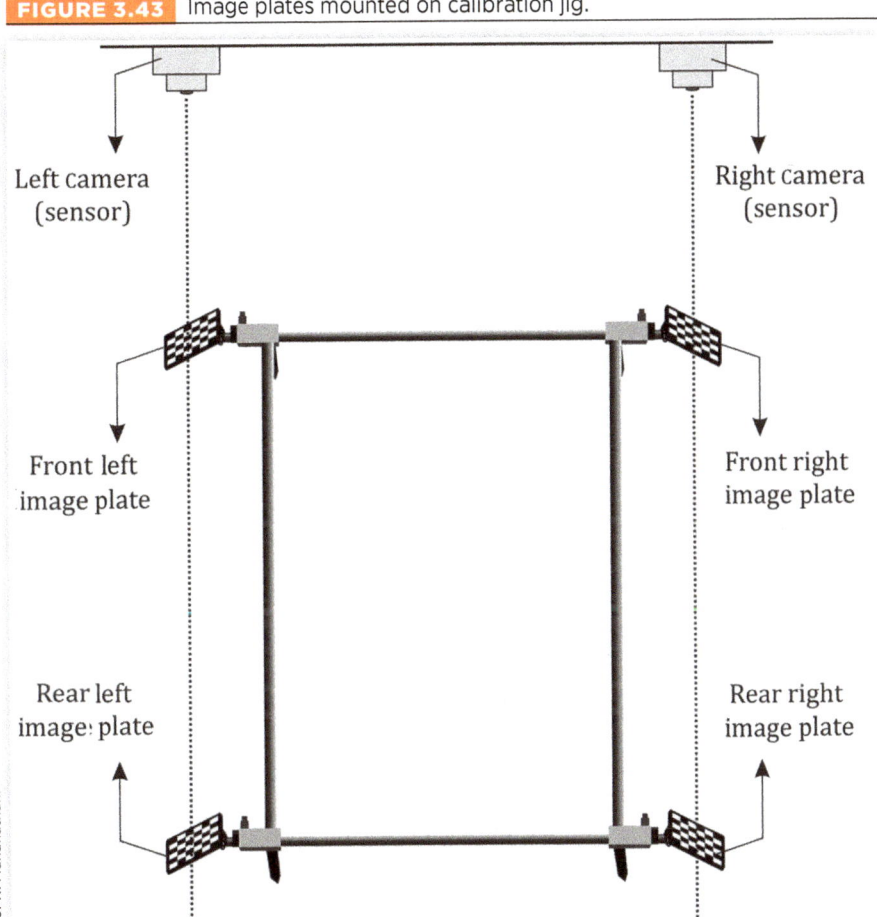

Now the right-side camera will see the right-side image plates (front and back) and the left-side camera will see the left-side image plates (front and back).

The image plates used in the four locations must be the same image plate **ear marked** for that particular location by the manufacturer. The location means Front Left, Front Right, Back Left, and Back Right. These image plates must be mounted to their respective places in the Calibration Kit. Using the small spirit level available in the image plates, they must be levelled and firmly secured to the kit.

Now all the four image plates will have their X, Y, Z angles in zero condition. This means the twelve angles of the four image plates in the three X, Y, Z planes are set to zero. Now this **zero** condition of the plates will be sensed by the camera through their images.

Since the sensors in the cameras are not calibrated, they may not display **zero** in all twelve planes. They will display values with minor variations depending on the field conditions. At this point in time, if the CALIB button in the system is pressed, the aligner software will make all the twelve readings display **Zero** values. By this process, the Field Calibration of the aligner is completed.

Now it is understood that the activity of sensing the zero condition of the image plates, in all the three planes, and making the camera display zero values is called **Calibration**. In other words,

The activity of sensing the "set values" of a sensor and making the system display the same "set values" is called "Calibration."

Subsequent to calibration all the measurements made and the readings displayed by the aligner will be correct in values.

Caution: The calibration procedure described above is a general explanation for understanding calibration. In practice, the procedure given for calibration in the Service or Installation Manual of the Wheel Aligner manufacturer must be followed meticulously, without any deviation.

After calibration, the Wheel Aligner is ready for operation. Once the calibration is done, normally it will not change unless any one of the following situations arises:

- In the event of the camera or the image plate is changed for service purposes.
- If the aligner is moved to a different location.
- If the aligner meets with an accident.

Note: There are some Wheel Aligners that do not require calibration after installation, as per the design. Such aligners must be installed strictly according to the installation procedure given in the Wheel Aligner manual.

3D Wheel Alignment Procedure

The procedure given in the User Manual by the supplier of Wheel Alignment computer has to be followed for carrying out Wheel Alignment on vehicles. The procedure described below is a general procedure for 3D Wheel Aligners and explains the basic operations.

Step 1: Park the vehicle for Wheel Alignment

The vehicle to be aligned has to be slowly driven and parked in front of the Wheel Aligner as shown in **Figure 3.44**.

FIGURE 3.44 Vehicle parked in front of Wheel Aligner.

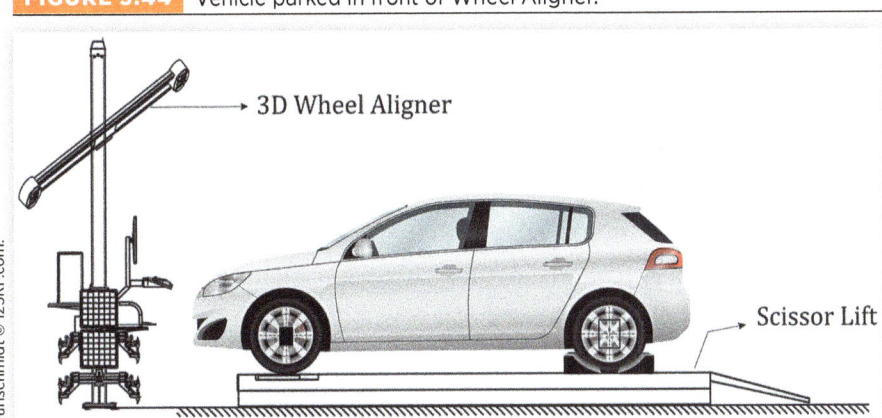

Courtesy of R. Mananathan; Vehicle: yurischmidt © 123RF.com.

If a pit is used instead of a lift, the vehicle has to be parked on the pit.

Step 2: Verify air pressure in all the wheels

Before starting Wheel Alignment, the air pressure in all four wheels must be checked. If the pressure is not correct, it must be rectified by filling air in the required wheels.

Step 3: Input vehicle data

The vehicle's details like registration number, owner name, kilometer run, etc. have to be given as input to the computer.

Step 4: Check and record the tire condition

Some Wheel Alignment computers give provision for recording the tire conditions before alignment. If so, it can be recorded in the system.

Step 5: Fix the wheel clamps

As explained earlier, the wheel clamps must be fixed to the wheels carefully in good locations and the clamping pins must be squarely seating on the wheel rims. In old vehicles, the wheel rim might have got distorted. It is **very important** to avoid such bent or distorted portions while fixing the wheel clamps.

Step 6: Fix the image plates

Fix the respective Image Plates to the wheel clamps as shown in **Figure 3.45**. Make it **true vertical** using the spirit level available in the Image Plates.

FIGURE 3.45 Image plate fixed to the wheel.

Step 7: Steering wheel straight ahead

After alignment, when the vehicle travels on the road in a straight path, the steering wheel must look geometrically straight as shown in Figure 3.46.

FIGURE 3.46 Steering wheel straight ahead.

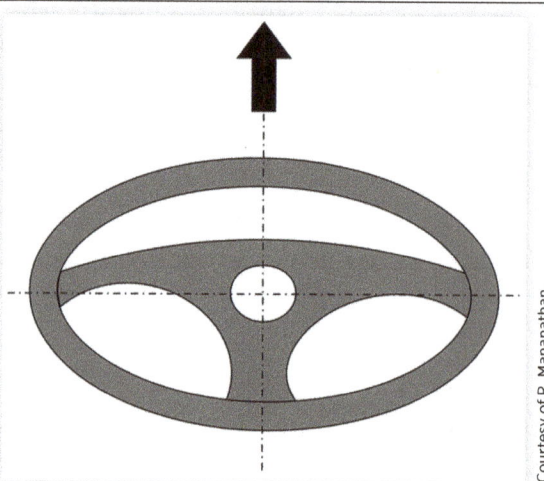

This geometrically or aesthetically looking straight position of the steering wheel is called **Steering Wheel Straight Ahead** position. During alignment, the technician has to turn and adjust the steering wheel and stop it in this position while measuring the initial readings and also while effecting corrections of the Wheel Alignment parameters. This is a very important condition in Wheel Alignment because all the Wheel Alignment parameters are measured in a condition at which the vehicle will travel in a straight-ahead position. This assumes importance in the following programs of Wheel Alignment:

- While carrying out **Run-out compensation**.
- While measuring all the **initial** parameters of Wheel Alignment.
- Finally, when angle **corrections** are made in the wheels.
- Particularly, while correcting the Toe angle because the Toe, which is set in straight-ahead condition, will be the correct Toe.

It is the duty of the technician to ensure the **Straight Ahead** condition carefully and correctly. Even though there are different types of steering wheels, the technician will normally know the correct **straight ahead** position of the steering wheel.

Step 8: Fix the steering wheel lock

During **Run-out** operation, when the vehicle is pushed by 30° or 180° in the forward and backward directions, the steering wheel should not get disturbed even slightly. For this, the **Steering Wheel Lock** is used. If the steering wheel lock is fitted properly, the steering wheel will get locked in the Straight-Ahead position and will not shake or move during the **Run-Out** measurement process.

Step 9: Wheel Run-out measurement

The first step in Wheel Alignment is to find out the value of run-out in all four wheels. For this, we have to push the vehicle back and forth to the extent of the wheels moving by 30° rotations. As per the pictorial guidance given by the computer, the vehicle has to be pushed **back** and then pushed **forth** slowly without jerks till its front wheels reach the rotary plates. The wheel stopper prevents the vehicle from crossing the limits while pushing. During the process, the camera acquires the X, Y, Z readings of all the four wheels, both at the **START** and at the **END** of pushing. These data are required to find out the **Run-out** value of all the wheels. The acquired X, Y, Z readings are sent to the computer by the camera. Using these data, the computer calculates the **Run-out** of all four wheels and keeps them in its memory. After the **Run-out** program the front wheels are seated on the rotary plates in **Straight Ahead** position, to proceed for further measurements.

Step 10: Fix the brake pedal lock

The next step is to measure the Caster and Kingpin angles. To carryout this, the steering wheel has to be turned by 10° towards the left side and then by 10° towards the right side from the **Straight Ahead** position. While turning the steering wheel, the vehicle should not move forward or backward. To ensure this, the Brake Pedal Lock has to be applied on the brake of the vehicle, which firmly keeps the brake pedal pressed, preventing any movement. This must be done before starting the Caster and Kingpin measurement process.

Step 11: Measuring Caster and Kingpin angle

After fixing the Brake Pedal Lock, the computer screen pictorially guides for turning the steering wheel by 10° towards the left side, STOP, and 10° towards the right side from its **Straight Ahead** position (Ref. Figures 3.47a, b, and c).

FIGURE 3.47a Move towards the left.

FIGURE 3.47b Move towards the right.

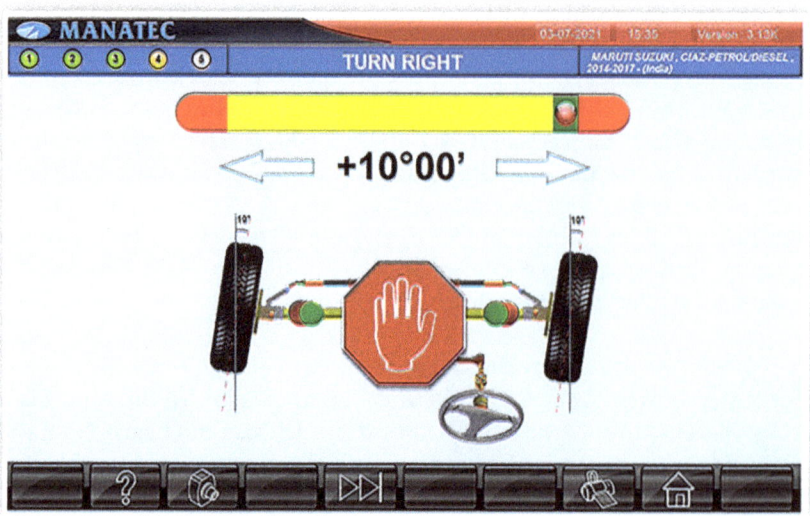

FIGURE 3.47c Stop at straight-ahead position.

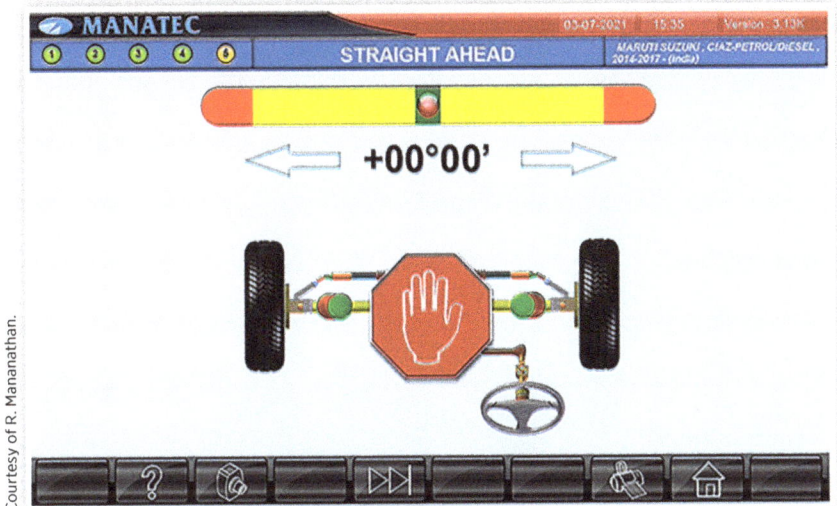

The mechanic must follow the guidance carefully and complete this **swing operation**. (In some Wheel Alignment computers, a 20° swing is followed.) The camera acquires the X, Y readings at the end of the left swing and at the end of the right swing positions and transmits the data to the computer for calculating the Caster and Kingpin angles. The calculated values are saved into the computer.

After the swings, the steering wheel will be brought back to the straight-ahead position by the screen guidance.

Step 12: Initial alignment readings

Now the computer has calculated the initial condition of all the Wheel Alignment parameters and displays them on the computer screen as shown in Figure 3.48.

FIGURE 3.48 Initial alignment readings.

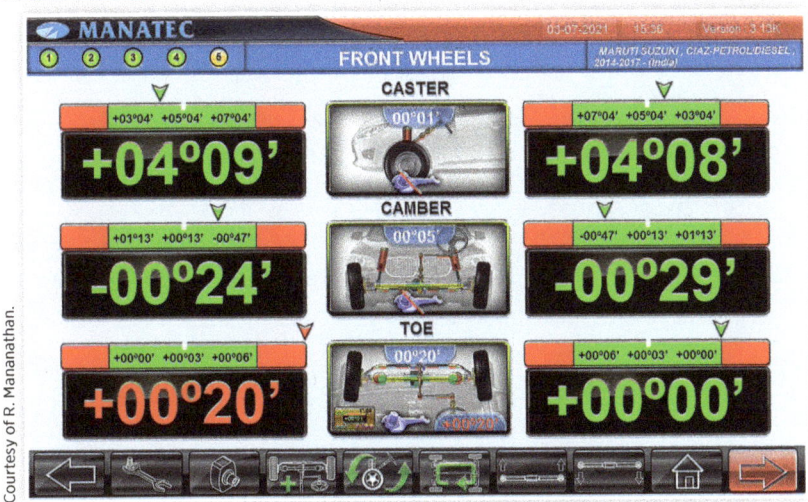

These values displayed are **Run-out** compensated and **Thrust angle** compensated, and these values indicate the initial alignment condition of the vehicle before carrying out any correction.

4

Wheel Alignment Correction Methods

Contents

Wheel Alignment—Types of Axles 69
Wheel Alignment—Correction Methods 72

Wheel Alignment—Types of Axles

Before proceeding to know the correction methods, it is important to know the basic design of axles with respect to Wheel Alignment.

In a vehicle, the part to which the wheels are mounted is called axle. There are many different types of axles, and they are named and called depending on their applications and design.

For Wheel Alignment purposes, three types of axle designs can be defined, and it is important to know about these three designs:

1. Rigid Axle.
2. Axle with Single Tie Rod.
3. Axle with Stub Axles.

Rigid Axle

FIGURE 4.1 Rigid axle.

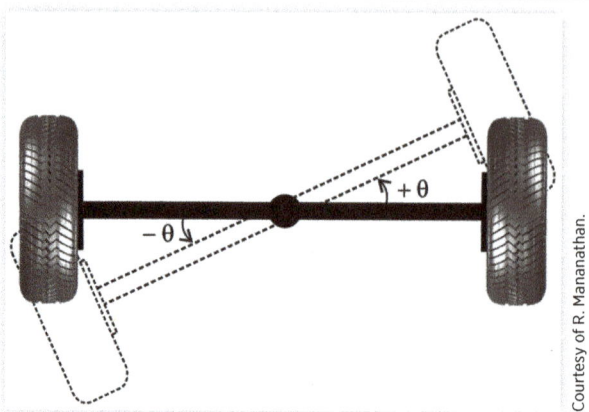

In a Rigid Axle, the left and right wheels are rigidly mounted on the axle, and for this reason, if the wheels have to be turned in the horizontal plane, the entire axle has to be turned as shown in Figure 4.1. In this case, if the right side is tilted by $+\theta'$, the left side will get tilted by $-\theta'$. Normally, the **rear axle** of LCVs and some axles in the HCVs will be rigid axles.

Axle with Single Tie Rod

FIGURE 4.2 Axle with single tie rod.

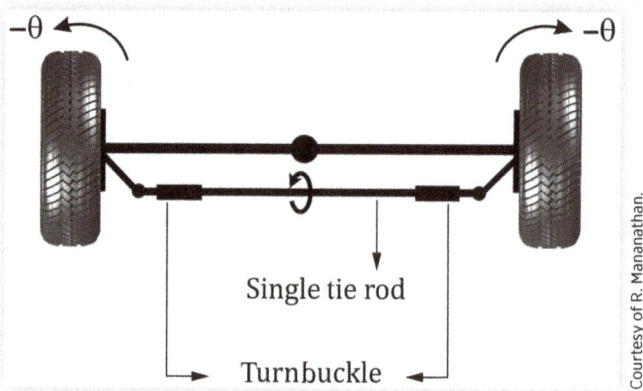

In this type of axle (Figure 4.2), the left and right wheels will be connected by a Single Tie Rod. Since the tie rods have opposite threads on the left and right sides (turnbuckle), when the tie rod is rotated, the left and right wheels will get tilted simultaneously either inwards or outwards. This means if the right wheel gets tilted by $(-\theta')$, the left wheel will also get tilted by $(-\theta')$. This kind of design is generally seen in HCVs.

Axle with Stub Axles

FIGURE 4.3 Axle with stub axles.

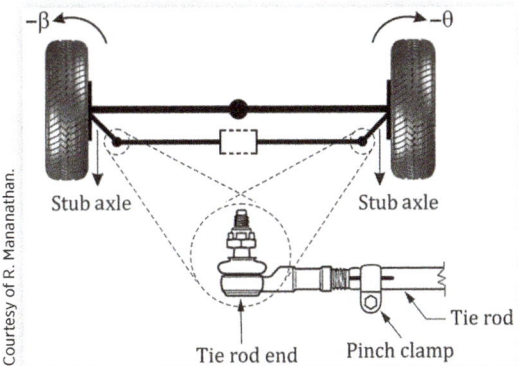

In this design, the tie rod is connected to the left wheel and right wheel through a stub axle. It is possible to change the Toe angle of the wheels independently by adjusting the stub axle. If the right wheel is adjusted to $(-\theta')$, the left wheel can be adjusted to $(-\beta')$. At the connecting point of the tie rod and the stub axle, there is a threaded mechanism, as shown in Figure 4.3. Using this arrangement, the left and right wheels can be adjusted individually to get the desired Toe values. Normally, the front axle of LCVs will be of this design.

Drive Axle—Definition

FIGURE 4.4 Drive axle.

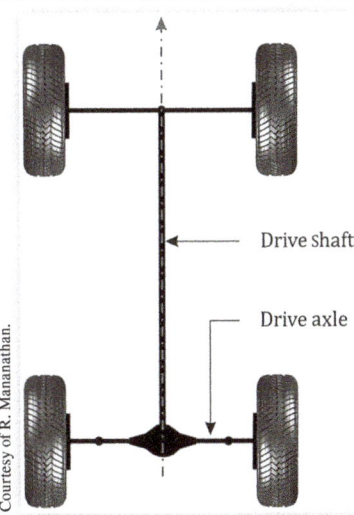

Drive axle is not a type of axle as described above. A drive axle is an axle that is driven by the engine through a **driveshaft**, as shown in Figure 4.4. The other axles are pushed and driven by the pushing movement of the drive axle. Normally, in LCVs, the rear axle will be the drive axle, though in some vehicles the front axle is designed as a drive axle.

Note: The axle figures shown above are only indicative line sketches. In practice, they look different.

Wheel Alignment—Correction Methods

Whenever vehicles come for alignment, mostly the Wheel Alignment angles will be deviating from the specifications or from the values set during the previous alignment session, for the following reasons.

When the vehicle runs on the road, depending on the road conditions, the vehicle is subjected to abnormal dynamic forces, which may impact the wheel rims, ball joints, bush bearings, suspension, etc.

The driving habit of the driver, like applying brakes, negotiating curves, etc., leads to a similar situation.

Even the normal wear and tear may not be uniform in the various joints and bearings of the suspension system that affect the wheel position in the axle assembly.

Since the Wheel Alignment specifications are in minutes accuracy, even a small impact or wear and tear affect the Wheel Alignment parameters.

Once the deviation is measured and given by the Wheel Alignment computer, correction must be done manually to bring them back to specified values.

Every vehicle manufacturer gives the specification of the Wheel Alignment angles for all the vehicles and models manufactured by them. Normally in the specifications, each parameter will have a **minimum value** and a **maximum value** as limits. On measuring the **initial condition** of the Wheel Alignment parameters in a vehicle, the parameters that are out of the limit alone need to be corrected to complete the Wheel Alignment. The correction procedure for a particular parameter may not be the same for all the vehicles. It may vary depending on the make and model of the vehicle.

The Wheel Alignment computer manufacturers will offer training to the technicians not only on using the Wheel Aligners but also on effecting the corrections of Camber, Toe, and Caster angles as applicable for different vehicles.

Color Display

The initial values of the angles are displayed **live** on the screen in two colors, i.e., green and red. Green means the angle is within limits and no further correction is required. Red means the angle is out of the limits. This color display helps the technicians to understand the present condition easily and helps in effecting the corrections quickly. Most of the Wheel Alignment computers follow this color code.

When a particular angle is in red color, i.e., out of the limits, while effecting the correction, the red color bar moves towards the specification, and on reaching **within the limits**, the color changes to green. As soon as the green color is achieved, the technician can stop further corrections and firmly tighten the nuts to complete the operation.

Once the red color parameters are corrected and brought to a green color, all the angles will be displayed in green color, except for the Kingpin angle, which cannot be corrected. These final angles are saved in the Wheel Alignment database, and if needed, a printout can be taken.

Though the correction procedure given in this book may be appropriate for most vehicles, the method of corrections for any particular vehicle has to be chosen from the service manual of the vehicle.

Apply Steering Wheel Lock

Earlier it was mentioned that the angle corrections must be done, keeping the steering wheel in a straight-ahead position. To ensure this, the steering wheel lock has to be applied firmly keeping the steering wheel in a straight-ahead position before starting angle corrections.

Toe Correction

Normally, in LCVs, there will be a tie rod having an adjustable threaded mechanism at the ends. The left and right toe values can be corrected individually to get the correct toe. See Figure 4.5.

FIGURE 4.5 Typical tie rod.

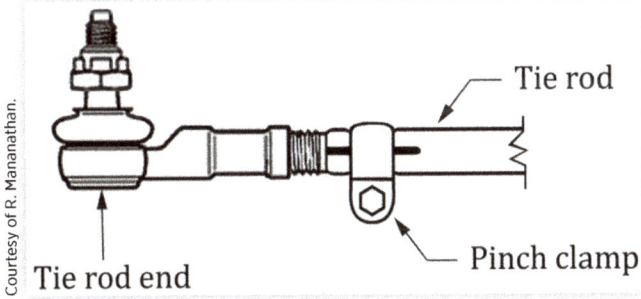

The following steps can be followed for effecting the corrections:

1. Loosen the pinch clamp bolt and nut.
2. Rotate the adjustment nut using a spanner and bring the correct Toe by observing the screen.
3. Again, tighten the lock nut firmly.

During this operation, the tie rod alone must be rotated carefully without rotating other parts.

 Important: If there is a provision in the vehicle for adjusting the rear wheels' toe, the rear wheel toe must be adjusted first, and then only the front wheel adjustment has to be carried out.

Cam Type Adjustment for Toe

In some vehicles, instead of a threaded mechanism, **cams** are provided for adjustments. A typical cam type is shown below in **Figure 4.6**.

FIGURE 4.6 Cam adjustment for Toe.

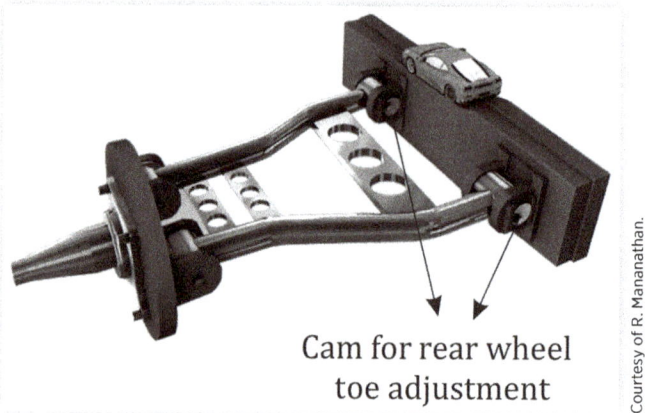

Cam for rear wheel toe adjustment

Two cams are provided for each wheel, one on the front side and the other on the backside of the wheel as shown in the figure. By adjusting the cams suitably, the toe angle of the wheel can be varied.

Camber Correction

The camber must be adjusted by following the procedure given in the service manual of the respective vehicle. However, a general procedure is given below.

Cam Type Adjustment for Camber

FIGURE 4.7 Cam adjustment for camber.

Cam for camber adjustment

By adjusting the cam shown in **Figure 4.7**, the Camber angle can be changed to the desired value.

Camber Adjustment Using Link Rod

FIGURE 4.8 Link rod adjustment for camber.

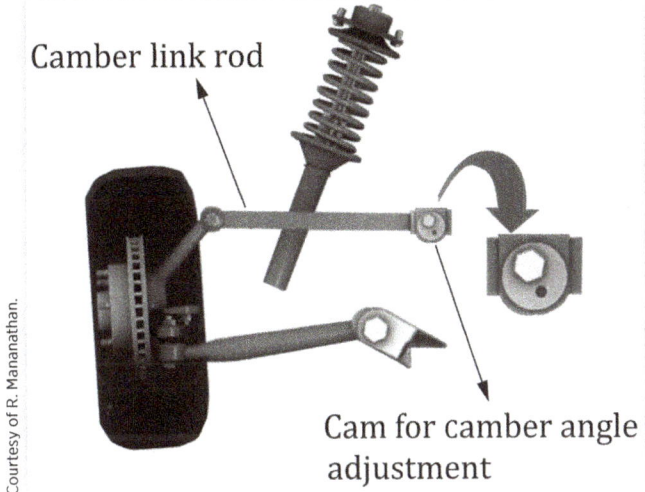

By adjusting the link rods shown in Figure 4.8, the Camber angle can be changed to the desired values.

Camber Adjustment Using Shims

FIGURE 4.9 Shim adjustment for camber.

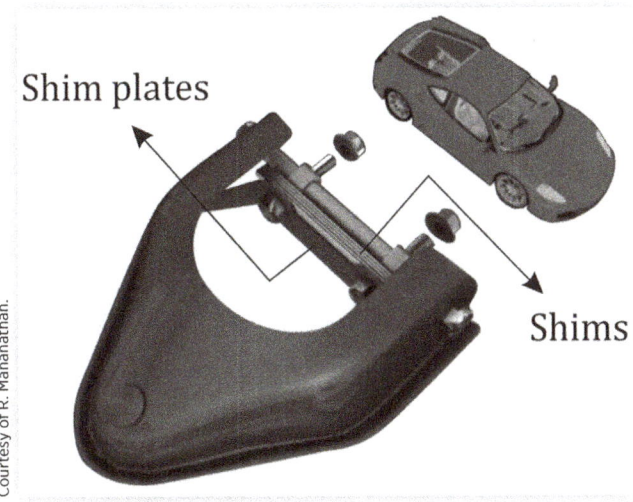

There are many vehicles in which shims can be added or removed to change the Camber as shown in Figure 4.9.

Caster Corrections

In most vehicles, the Caster angle is a fixed one and cannot be adjusted. In some designs, the Caster can be adjusted. In these cases, the manufacturer would have given a specific procedure for adjusting the Caster, and this procedure must be followed. However, a general example is given below in Figures 4.10 and 4.11.

Caster Adjustment Using Cams

FIGURE 4.10 Cam adjustment for caster.

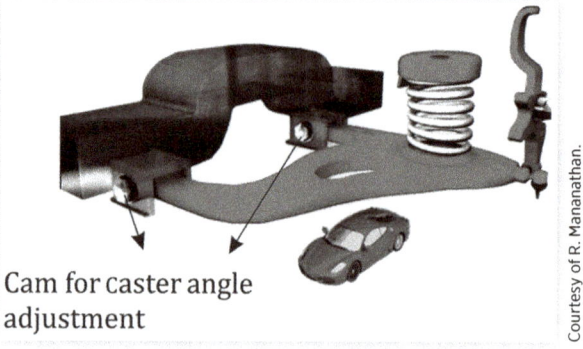

Caster Adjustment Using Cam and Link Rod

FIGURE 4.11 Cam and link adjustment for toe.

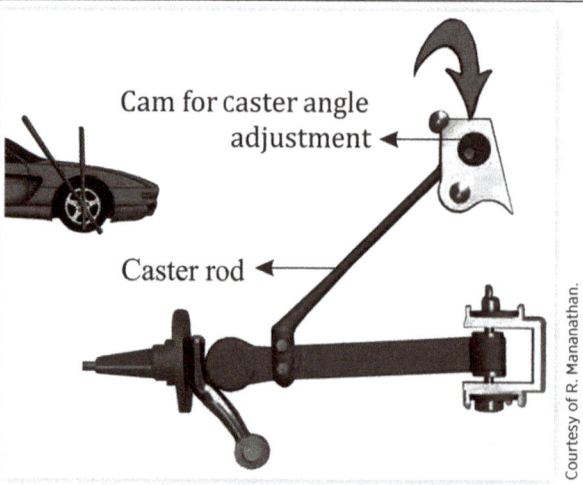

Note: All the above examples are from generally followed designs. But the technicians are advised to study the Service Manual of the respective vehicle and follow the procedure given in the manual. It is important for the technicians to take proper training before effecting corrections.

After correcting all the required angles, the final angles are shown in the display. A printout can now be taken. A sample printout is given below in **Figure 4.12**.

FIGURE 4.12 Wheel Alignment printout.

WHEEL ALIGNMENT CENTRE
Mohan Nagar,
Puducherry.

WHEEL ALIGNMENT RESULTS   Job Number - 036

Date / Time : 11-Dec-17 6:15 pm		Regn. No.	: PY-01-AB-2856
Mechanic : Vimal		Owner	: Manatec
Odometer : 45955		Address	: Industrial Estate
Make : Maruti Suzuki		City	: Puducherry
Model : SX4		State	: Puducherry
		Phone	: 0413 2248926

DB : India (2014-1)	Before correction	Specification			After correction
🟩 **Within limits**		Min	Target	Max	
🟧 **Out of limits**	**FRONT WHEEL ALIGNMENT RESULTS**				
CAMBER Left	−00°14'	−00°22'	+00°38'	+01°38'	−00°14'
Right	−00°25'	−00°22'	+00°38'	+01°38'	−00°26'
Max.diff. Left/Right	00°11'		01°00'		00°12'
KINGPIN Left	+12°08'	+11°56	+11°56'	+11°56'	+12°08'
Right	+12°28'	+11°56'	+11°56'	+11°56'	+12°08'
Max.diff. Left/Right	00°20'		00°00'		00°20'
INCLUDED ANGLE Left	+11°54'	---	+12°34'	---	+11°54'
Right	+12°03'	---	+12°34'	---	+12°02'
Max.diff. Left/Right	00°09'		---		00°08'
CASTER Left	+03°46'	+02°30'	+03°30'	+04°30'	+03°47'
Right	+03°35'	+02°30'	+03°30'	+04°30'	+03°35'
Max.diff. Left/Right	00°11'		01°00'		00°12'
SETBACK	10.5mm	---	---	---	10.5mm
TOE Left	+00°03'	+00°00'	+00°04'	+00°09'	+00°05'
Right	+00°11'	+00°00'	+00°04'	+00°09'	+00°08'
Max.diff. Left/Right	00°08'		00°04'		00°04'
TOTAL TOE	+00°14'	+00°00'	+00°09'	+00°18'	+00°14'
	REAR WHEEL ALIGNMENT RESULTS				
CAMBER Left	−01°20'	+00°00'	+01°00'	+02°00'	−01°20'
Right	−00°41'	+00°00'	+01°00'	+02°00'	−00°41'
Max.diff. Left/Right	00°39'		01°00'		00°39'
THRUST ANGLE	−00°02'	---	---	---	−00°04'
TOE Left	+00°06'	+00°00'	+00°21'	+00°42'	+00°06'
Right	+00°11'	+00°00'	+00°21'	+00°42'	+00°15'
Max.diff. Left/Right	00°05'		00°21'		00°09'
TOTAL TOE	+00°17'	+00°00'	+00°42'	+01°24'	+00°21'
SETBACK	11.5mm	---	---	---	11.5mm

Fox 3D (S/W : 3.06, S/N :3D3230)   Copyright © 2010, Manatec Electronics

Courtesy of R. Mananathan.

In this Printout, the specifications of all parameters, the values before alignment and after alignment, are given in detail. If the final alignment readings are out of specification, such readings are printed in red color.

This Printout can be given to the vehicle owner, and a copy can be retained by the alignment center for records.

5

Wheel Alignment of Heavy Vehicles and Trailers

Contents

Heavy Commercial Vehicles—An Overview	79
Axle Configurations in HCVs	80
Types of Axles in HCVs	81
Wheel Alignment Parameters in HCVs	83
HCV Aligner Description	88
Infrastructure Required for HCV Alignment	89
HCV Aligner—Field Calibration	91
HCV Alignment—Procedure	92
HCV Angle Corrections	96
Chassis or Frame Alignment	103
Trailer Alignment	106

Heavy Commercial Vehicles—An Overview

FIGURE 5.1 Multi-axle truck for alignment.

Courtesy of R. Mananathan; Dump Truck: Nerthuz/shutterstock.com.

©2022 SAE International

The following types of vehicles are classified under HCVs:

- Buses.
- Trucks.
- Multi-axle trucks.
- Trailers.
- Articulate buses.

Though the alignment principle is the same for both LCVs and HCVs, the alignment of HCVs varies in many ways (Figure 5.1). Since HCVs carry heavy loads and are having many types of axle configurations, the complexity of the wheel geometry, its parameters, and the procedure of alignment are different from LCVs. The alignment pit and accessories needed for the HCV alignment are much bigger in size and are heavy. Since the HCVs and trailers have a very long wheelbase, powerful cameras are required to view the long-distance wheels and their image plates. Also the number of cameras required are more. The efforts demanded from the technicians to carry out alignment are more compared to the LCV alignment. For these reasons, the HCV aligners are costlier than LCV aligners.

The chassis of HCVs are larger in size. Therefore, the axles used in the chassis are longer in size. For this reason, while manufacturing HCVs, certain distortions in the chassis and axles are inevitable. A good Wheel Alignment computer can compensate for the effects of such distortions and make the vehicle run on its driveline. Particularly in multi-axle vehicles, even a small variation in alignment will wear out the tires rapidly within a few thousand kilometers of run. Therefore, Wheel Alignment is very critical and important for multi-axle vehicles.

The cost of HCV tires is very high. Periodic Wheel Alignment will reduce the tire wear, and substantial cost can be saved. If the maintenance expenses of a HCV is calculated for one year, the tire cost will be the maximum. Therefore, the expenses incurred for regular Wheel Alignment will be found reasonable, considering the savings derived from such Wheel Alignments. Investing a high cost on HCV Wheel Aligners is also justified.

Axle Configurations in HCVs

In HCVs there are many types of axles, and they are named and called depending on their design and applications. The following configurations are normally available in multi-axle HCVs:

- Buses and ordinary trucks have two axles similar to light vehicles.
- In trucks, there are vehicles having more than two axles. These are called multi-axle vehicles.
- All the multi-axle vehicles do not have similar configurations; they vary as shown in Figure 5.2.

FIGURE 5.2 Different axle configurations in HCVs.

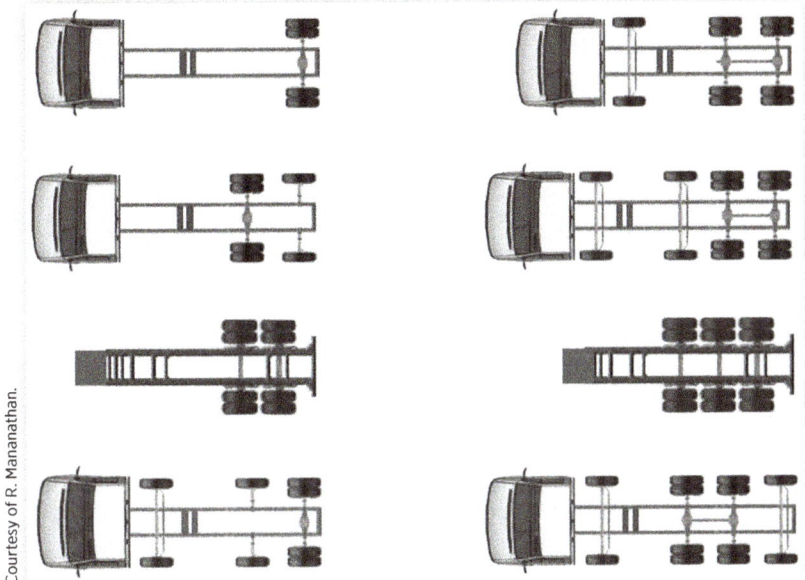

- Besides what is shown in Figure 5.2, there are many more configurations.
- There are axles having two wheels per side amounting to four wheels in an axle based on the load distribution design.
- There are vehicles measuring up to 19.0 meters long.
- Trailers and articulated buses are also very long and have many different axle configurations.

Types of Axles in HCVs

To understand the various axles in a heavy vehicle, a six-axle vehicle is shown in Figure 5.3.

Front Axle (F1): In buses and trucks, normally one Front Axle will be there (F1).

Twin Steer Axles (F1, F2): In some multi-axle vehicles, there are two Front Axles, and both can be steered. They are called Twin Steer Axles. These two axles are interconnected by a Drag Link as shown in Figure 5.4. They steer together while running. The axles F1 and F2 have a single tie rod, as shown in the figure. If this tie rod is rotated, both the left wheel and right wheel will turn simultaneously either towards the inside or towards the outside.

Drive Axle (B3): Drive Axle is the axle that is driven by the engine directly through a Driveshaft. The other axles are pulled or pushed by the movement of the drive axle. Irrespective of the number of axles in an HCV, the alignment of all the wheels must be carried out with respect to the Drive Axle only. This means all other

82 Automobile Wheel Alignment and Wheel Balancing

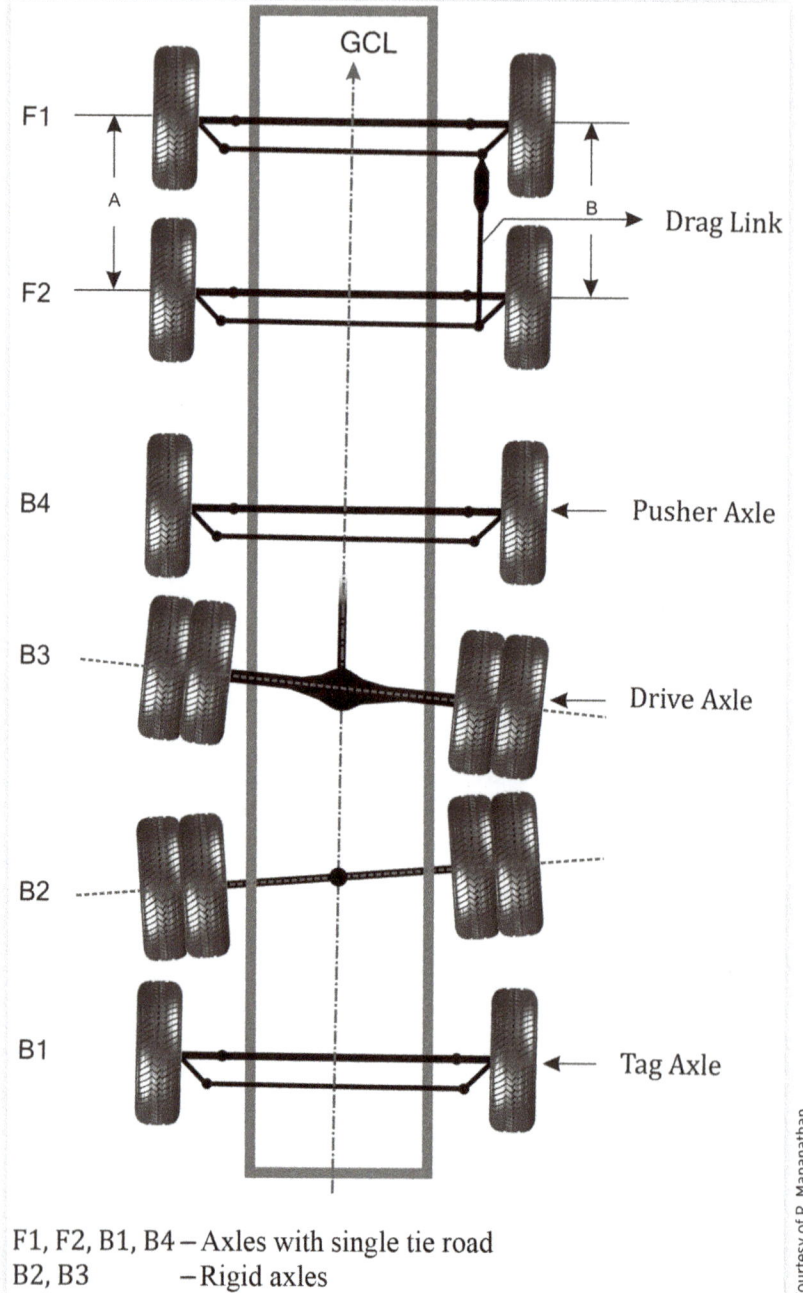

FIGURE 5.3 Six axle chassis with wheels.

F1, F2, B1, B4 – Axles with single tie road
B2, B3 — Rigid axles

axle's Thrust Line (Driveline) must be aligned with the Thrust Line of the Drive Axle. Therefore, while aligning a multi-axle vehicle, it is important to know the Drive Axle.

Generally, the Drive Axle is located behind F1 axle in vehicles having only one Front Axle and behind the F2 in vehicles having Twin Steer Axles.

FIGURE 5.4 Twin Steer Axles with drag link.

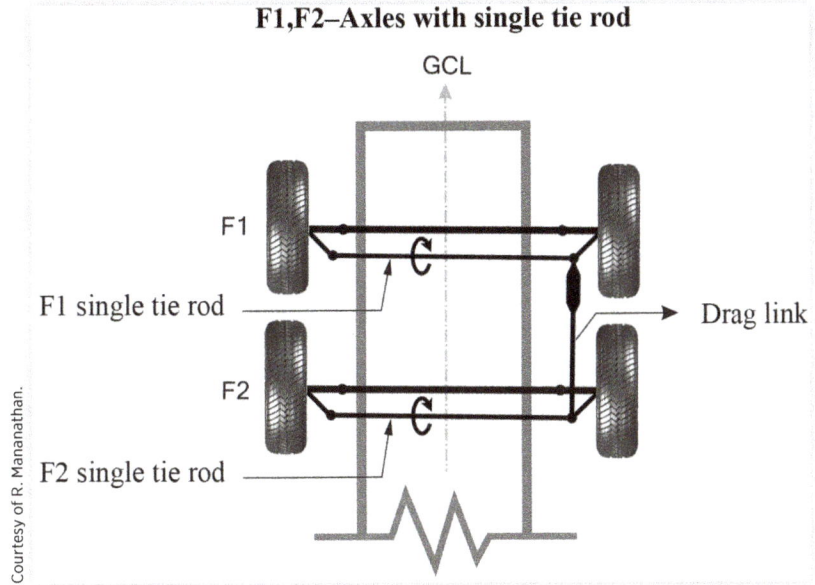

 Note: In multi-axle vehicles, there are designs having more than one drive axle. In such cases, the Wheel Alignment must be carried out with respect to the first axle, which gets power from the engine.

Other Axles (B2): Other than the drive axle, there can be more axles depending on the vehicle design. In **Figure 5.3**, such an axle is shown as B2.

Pusher Axle and Tag Axle (B1, B4): In some vehicles, there are axles called Pusher Axle and Tag Axle. These are designed to touch the ground only when the vehicle is loaded beyond certain limits. Otherwise, they will remain in lifted condition and act as idle axles.

Trailer Axles: In trailers there are different types of axles. Normally, all the axles of Trailers will be Rigid Axles with wheels rigidly fixed to the ends.

Wheel Alignment Parameters in HCVs

All the Wheel Alignment parameters specified for LCVs are also applicable for HCVs. The definitions and explanations given in the **Light Vehicles** chapter for Camber, Toe, Caster, Kingpin, Run-Out, and Thrust Angles are applicable for HCVs also. Still some additional information are given below.

Camber

In HCVs, Camber is applicable for all the wheels of the multi-axles.

Toe

Toe angle is also applicable for all the wheels. In the case of rigid axles, **total toe** is considered relevant. The reason being, when a rigid axle is tilted for adjustment, both the left and right wheels will tilt together, keeping the total toe constant.

Caster

In HCVs, there may be one (F1) Front Axle or Twin Axles (F1, F2). In either case, caster is applicable separately for both F1 and F2 axles. Caster can be measured during Wheel Alignment but can be corrected in the workshop only and not in Wheel Alignment Centers.

Kingpin Angle

Kingpin angle is also applicable for both F1 and F2 axles. Kingpin angle will be measured during Wheel Alignment and can be corrected in the workshops only.

Thrust Angle and Thrust Line

The angle between the GCL and the Thrust Line is called the Thrust Angle (THA). Normally, the THA must be zero. This will be achieved by tilting and adjusting the drive axle.

The bisector of the drive axle's **Left wheel Toe** and the **Right wheel Toe** is called **Thrust Line**. If the left toe and the right toe of a drive axle are equal, then the Thrust Line will coincide with the GCL. Otherwise, the Thrust Line will separately exist in the vehicle geometry.

Wheel Run-Out

Wheel Run-Out has to be measured for all the wheels of a multi-axle vehicle for **Run-out compensation**. Since the HCVs have larger wheels and also have big size bearings, the run-out in HCVs will be more than the run-out of LCVs.

Geometric Centerline

GCL is the line drawn perpendicular to the front axle (F1) Through its middle point. All other axle's middle point must lie on this line only. Normally, the Toe angles are measured with respect to GCL. If there is a Thrust Line, the toe will be measured with respect to the Thrust Line.

Scrub Angle

As shown in Figure 5.5, the angle between the Thrust Line of the B3 axle and the Thrust Line of the B2 axle (bisector of left and right toes) is called Scrub Angle.

$$\text{Scrub Angle} = \text{B2THA} - \text{B3THA}$$

Automobile Wheel Alignment and Wheel Balancing   85

FIGURE 5.5 Six axle chassis with important parameters.

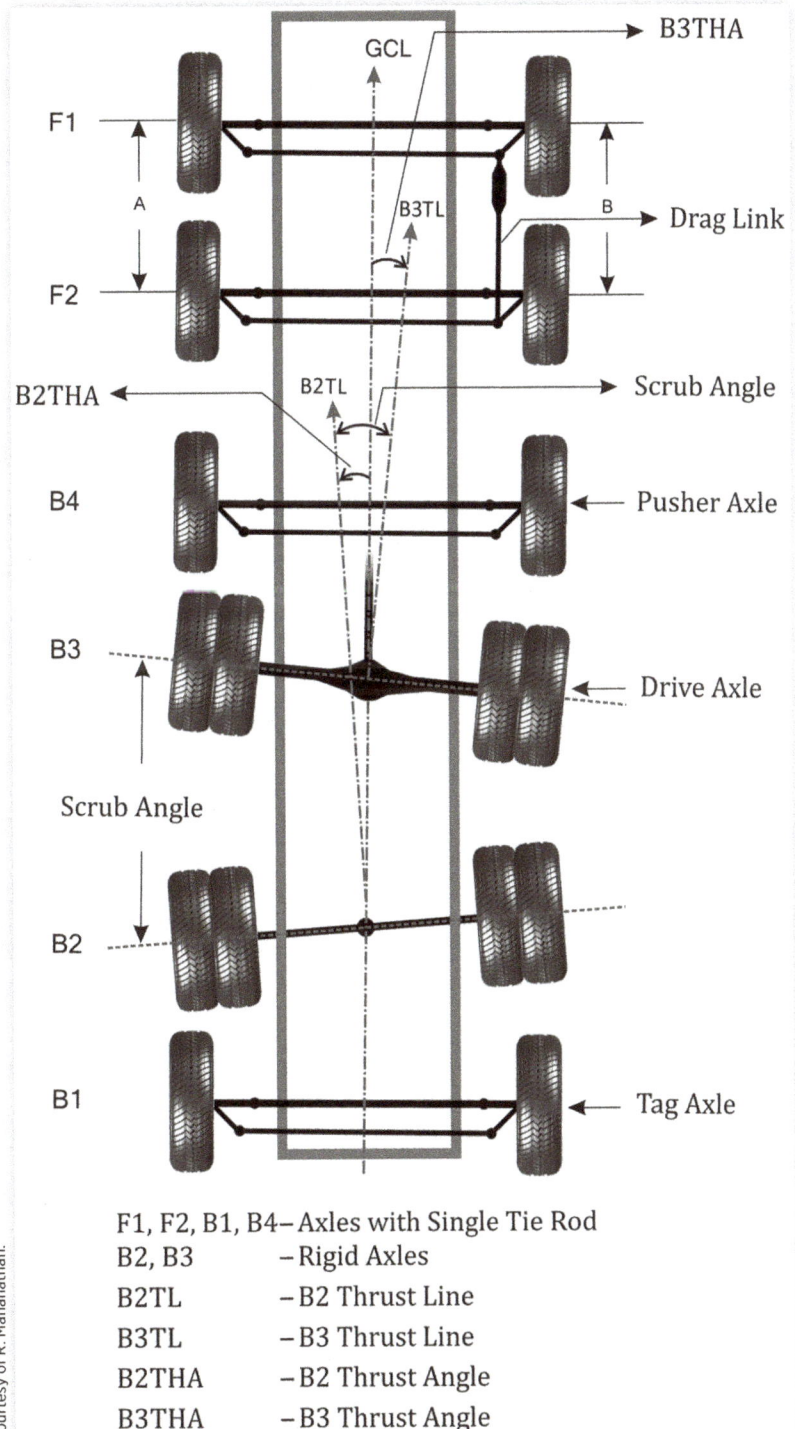

F1, F2, B1, B4 – Axles with Single Tie Rod
B2, B3 — Rigid Axles
B2TL — B2 Thrust Line
B3TL — B3 Thrust Line
B2THA — B2 Thrust Angle
B3THA — B3 Thrust Angle

Courtesy of R. Mananathan.

It is important to ensure that the Scrub Angle is **zero** in a vehicle. Otherwise, abnormal tire wear will take place.

Parallelism

FIGURE 5.6 Parallelism illustration.

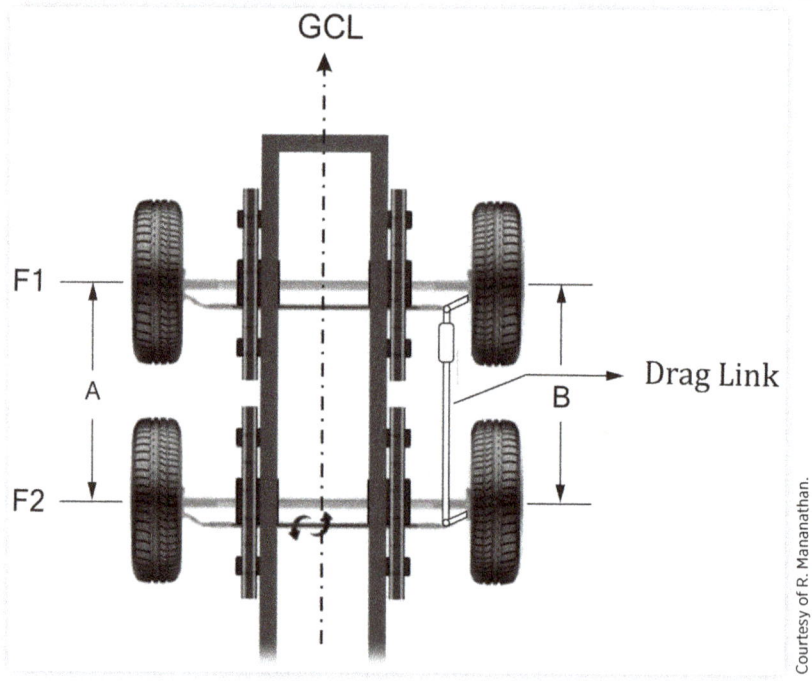

Parallelism is applicable only for Twin Steer Axle vehicles (Figure 5.6). The F1 and F2 axles must be parallel to each other. In other words, the difference between F1THA and F2THA must be zero. This condition is called **Parallelism**.

$$\text{Parallelism} = \text{F2THA} - \text{F1THA} = 0$$

If parallelism is not equal to zero, there will be abnormal tire wear in the F2 wheels. Before the introduction of computerized Wheel Alignment systems, parallelism was ensured in the following manner.

As shown in Figure 5.3, make A = B by tilting the F2 axle.

At the same time make the Left toe equal to the Right toe.

$$\text{F1LTO} = \text{F1RTO and F2LTO} = \text{F2RTO}$$

If the above conditions are achieved, Parallelism is achieved.

 Note: Other than the drive axle bisector line, the bisector lines of all other axles are not supposed to be called as Thrust Line. Since it is in practice to call all the bisector lines as Thrust Line, it is followed here.

Axle Shift in HCVs

In HCVs the center point of all the axles must lie in the GCL. (GCL is the perpendicular bisector of the F1 Axle). But in practice, they may not lie in the same line of the GCL. The points might have shifted either to the right side or to the left side from the GCL, as shown in **Figure 5.7**. This defect happens while mounting the axles during production. Even if all the axles are parallel to each other, the Axle Shift may be present.

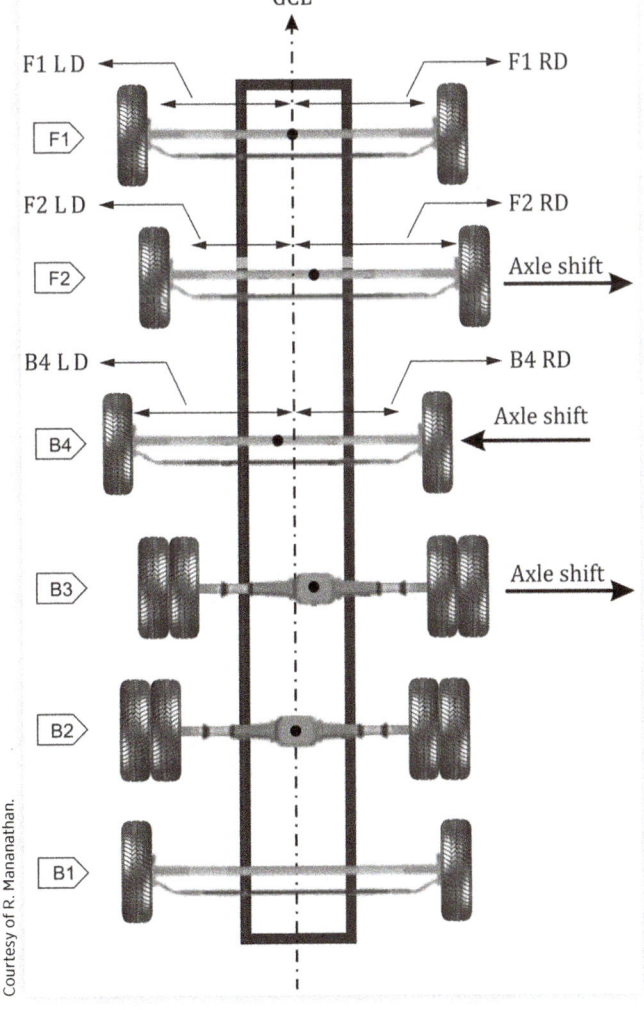

FIGURE 5.7 Illustration of an axle shift.

Such axle shifts do not affect the wheel alignment angles. Some 3D Wheel Aligners can measure the Axle Shift and will tell towards which direction the axle has shifted. The correction can be done only in the workshops.

HCV Aligner Description

Heavy Wheel Aligners are manufactured using CCD and 3D technologies. Since the CCD technology is vanishing, the installation of a typical 3D multi-axle Wheel Aligner is described below.

FIGURE 5.8 A typical HCV Wheel Aligner.

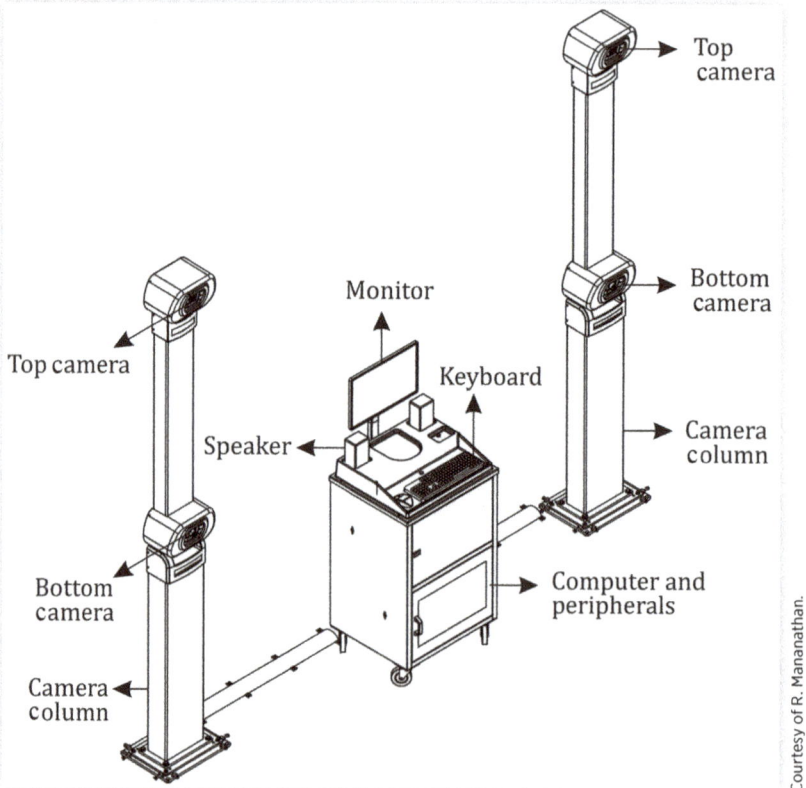

As shown in Figure 5.8, there are two Vertical Columns, one on the left side and the other on the right side of the pit, and located in front of the vehicle being parked for Wheel Alignment. These types of Wheel Aligners have two cameras per side, fixed on the two vertical columns.

There are models in which two cameras per side are fixed in a horizontal beam as shown in Figure 5.9.

There is a powerful computer available to receive all the images of the image plates fixed to the multi-axle wheels simultaneously and process them **at real time**. For this reason, the memory and speed of the computer will be very high.

The Rotary Plates and Wheel Clamps will be bigger in size and heavier in design to accommodate the HCV wheels. The Brake Pedal Lock and Steering Wheel Lock will be bigger in size.

FIGURE 5.9 A 3D HCV Wheel Aligner with cameras mounted on a horizontal beam.

Some special tools like Single Axle Pusher, Chain Block, Moving Trolleys, etc. will form part of the HCV Wheel Aligner.

Infrastructure Required for HCV Alignment

Many accessories are required to carry out HCV alignment successfully.

HCV Alignment Pit

There are many types of HCV Wheel Aligners available in the market. But the pit required for the technicians to go down and carry out angle adjustments will be the same for all types of aligners.

FIGURE 5.10 HCV alignment pit.

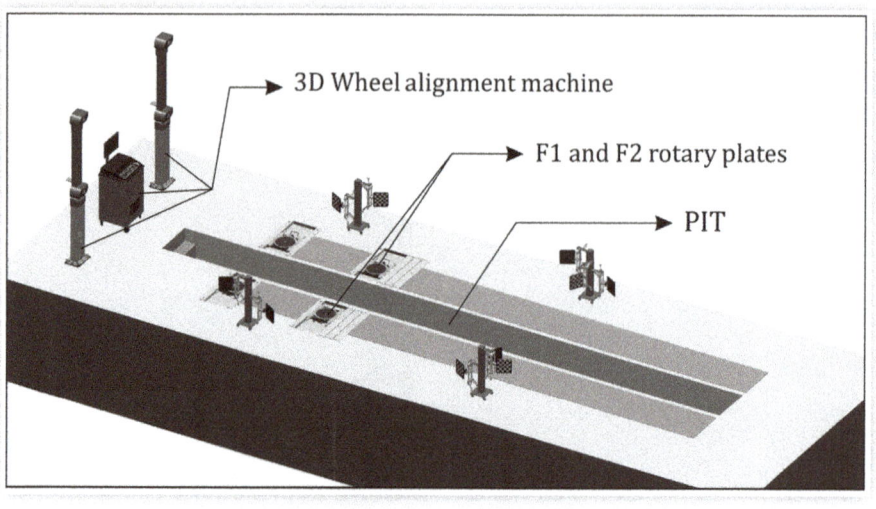

The pit seen in Figure 5.10 must have sufficient length to accommodate the longer trucks and sufficient depth for the technicians to comfortably go down and approach the axles and the tie rods for effecting corrections.

Two depressions are made per side to accommodate the Rotary Plates of F1 and F2 axles.

Except for the wheels resting on the Rotary Plates, all other wheels will be resting on the floor. This floor area must be a levelled surface, without any undulations. While constructing the pit, the surfaces must be carefully finished using either water level or spirit level. Since the HCVs are heavy in nature, the flooring must be a strong concrete floor with proper reinforcement. Along the edges of the pit, on either side, a rail must be embedded to run the moving trolley fitted with a jack.

If the **Run-out measurement** is going to be carried out by pushing the vehicle, the moving trolley may not be necessary. In 3D Technology, simultaneous run-out of all the wheels (up to six axles and twelve wheels) is also possible. In this case, instead of lifting the axle one by one, the driver can sit in his seat and move the vehicle by 30° to get the run-out of all the wheels at a time. This saves enormous time and efforts from the technicians. Some wheel Aligners use 180° movement.

Trolley Jack

The Trolley Jack as shown in Figure 5.11 is useful in the case of lifting each axle one by one for removing the wheels.

FIGURE 5.11 Trolley jack.

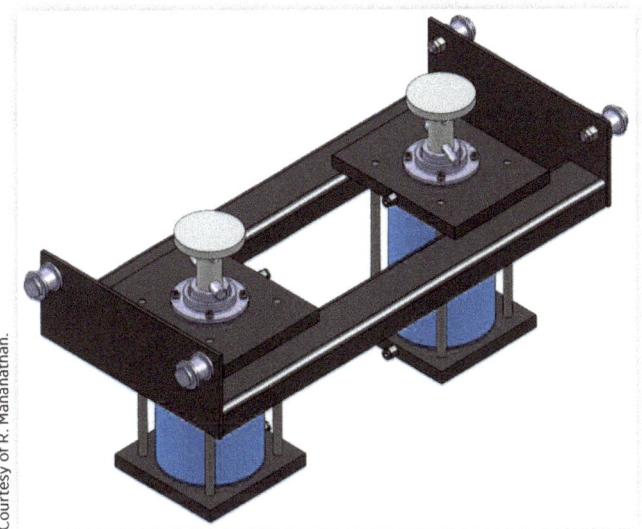

The jacks fitted in the trolley can be either a manual screw jack or pneumatically operated one.

Air Compressor

An air compressor is required for supplying compressed air to operate all the pneumatic tools.

HCV Aligner—Field Calibration

The principle of **Field Calibration** has already been explained in detail in the **Light Vehicles** chapter in this book.

Every camera available in a 3D Wheel Aligner must be calibrated in the field after installation. This Field Calibration is required because the installing environment like pit level, etc. in the field may not be the same as the environment that existed during the factory calibration.

The Field Calibration requires a Calibration Kit. This Kit will be a rectangular jig in which the Image Plates are mounted. The kit has to be levelled and positioned as per the **calibration procedure** given in the User Manual.

Now, if the cameras have been installed in perfect parallel condition to the image plates, the cameras are supposed to sense these values as zero and display them as zero. But in practice, while installing the camera columns, there may be minor variations that make the cameras read marginally different values other than zero.

By pressing the CALIB button, the cameras are made to display zero values by suitably modifying the values sensed by them. This activity is called Field Calibration.

After Field Calibration, the cameras or sensors will give the correct angular values during alignment.

Caution: Calibration is a critical activity. It must be carried out exactly as per the procedure given in the User Manual of the manufacturer. The procedure given in this book is an indicative general procedure only. This should not be taken as a calibration procedure for all the aligners.

Normally, the calibration done in Wheel Aligners will not change, unless the aligner meets with any accident, or the cameras are replaced for service reasons.

Note: There are certain designs that do not require Field Calibration. They are **self-calibrated** internally. For such types of Wheel Aligners, there is no need for Field Calibration. But the installation procedure must be strictly followed.

HCV Alignment—Procedure

In HCVs also, the Wheel Alignment must be carried out after balancing the wheels. For HCVs, alignment on the Lift is not applicable because of its heavy weight. The alignment must be carried out on pit only.

Parking the Vehicle for Alignment

Refer to Figures 5.12 and 5.12a. The vehicle must be slowly moved on the pit towards the camera and stopped when the front wheels are parked on the top of the F1 rotary plates.

FIGURE 5.12 Front wheel parked on F1 Rotary Plate.

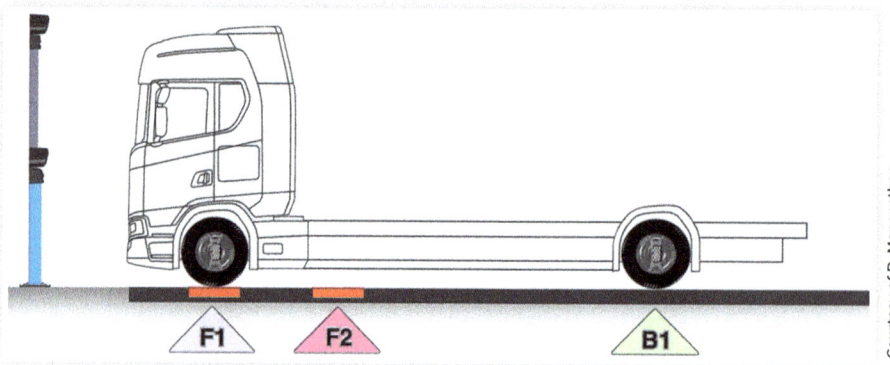

If there is an F2 axle, its wheels may not seat on the F2 rotary plate. Now the F2 Rotary Plate must be moved and placed below the F2 axle wheels. For this, the Rotary Plate pit will be longer in size, and suitable spacers are supplied by the Wheel Alignment manufacturers to hold the Rotary Plates in position.

For vehicles shown below, the front wheels will be parked in the F2 location to accommodate the balance of the three wheels within the purview of the cameras.

FIGURE 5.12A Front Wheel parked on F2 Rotary Plate.

 Note: The vehicle parking will be guided pictorially by the Wheel Alignment computer on the screen for different axle configurations. This must be followed.

Measurement of Wheel/s Run-Out

The procedure for HCV Wheel Run-out is the same as LCVs Run-out (Ref. Chapter 3). In LCVs, to perform run-out, the vehicle must be manually pushed by 30°. In the case of HCVs, instead of manual pushing, the vehicle can be moved by the driver by slowly driving the vehicle. (Some Wheel Aligner manufacturers recommend 180° travel for run-out.)

As explained earlier, the most important condition for the run-out program is to ensure that the steering wheel is kept in a **straight-ahead position** during the entire process of measuring run-out. For this, the steering lock has to be firmly fixed and locked before starting the Run out push. The Run out can be carried out either individually for every wheel or simultaneously for all the wheels depending on the design of the Wheel Alignment computer. Simultaneous run-out saves a lot of **time and effort** from the technicians.

Till the year 2017, the Run out of the wheels of multi-axle vehicles was carried out by jacking up the wheels one by one. This required more time and enormous human effort to raise the heavy axles using jacks. The 3D technology was not available for HCVs at that point in time.

 **Important Information:** In the year 2017, a new method was invented for simultaneous run-out measurement of all the wheels at a time up to six axles (twelve wheels). Not only the run-out but also all the Wheel Alignment parameters were acquired simultaneously after run-out and the caster swing operations. In this method, all the initial alignment readings were displayed within a few minutes. This invention not only saved a lot of time but also saved a lot of human efforts for the technicians and made the multi-axle vehicle Wheel Alignment a simple task.

Note: As soon as the run-out operation is completed, the computer keeps this data in its memory for all the wheels and uses it for compensating in the values of camber and toe angles.

Measurement of Caster and Kingpin Angles

After the run-out measurement, the front wheels F1 and F2 will be on their respective rotary plates and the steering wheel will be in the **straight ahead** position. Now the Caster and Kingpin measurement must be done. This requires the steering wheel to be turned to the left and right side by 10°. This action is called **caster swing**. During this **caster swing** process, the vehicle may roll on the rotary plates affecting the acquisition of data. To prevent this, the Brake Pedal Lock must be fixed firmly before starting the **caster swing** operation. The Brake Pedal Lock keeps the brake pressed throughout the operation.

Graphical Guidance for Caster Swing

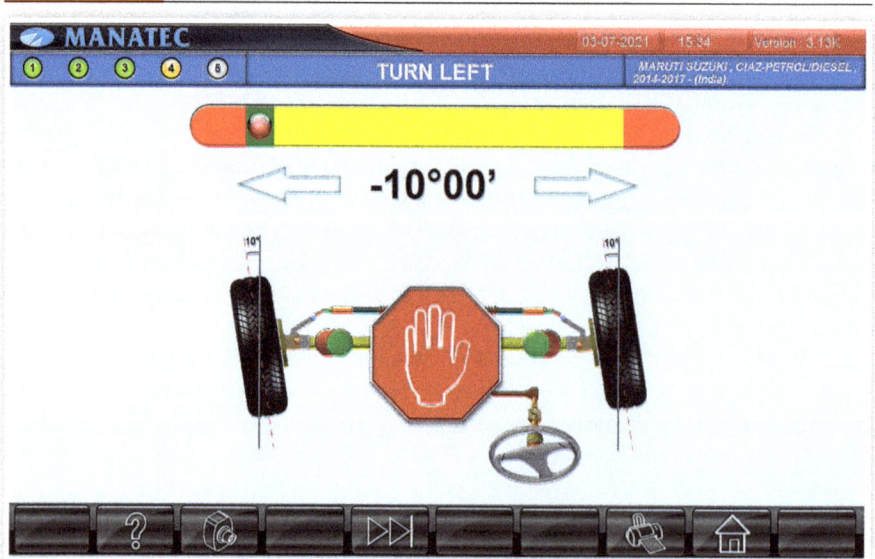

FIGURE 5.13 Turn left.

FIGURE 5.13A Turn right.

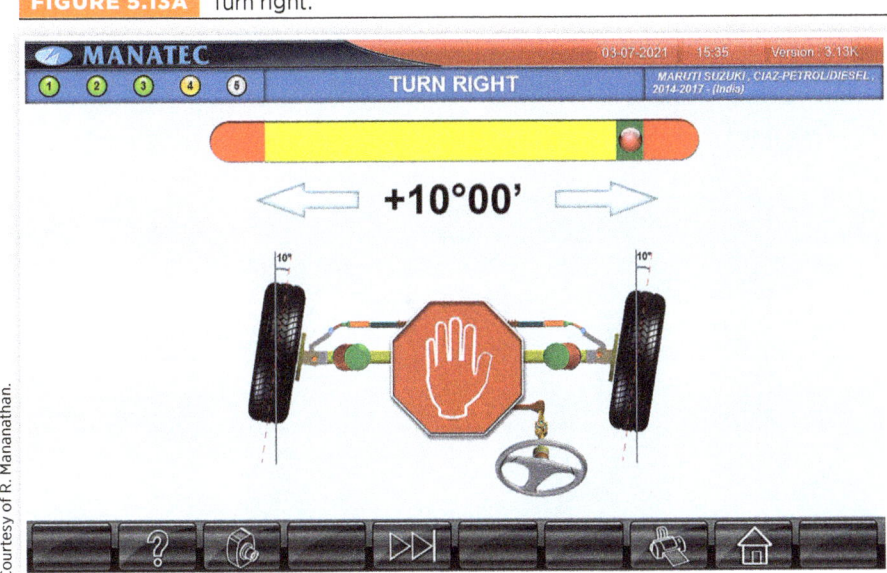

The technician must follow the graphical guidance on the screen as shown in Figures 5.13 and 5.13a. Slowly turn the steering wheel to the left side (−10°), wait for the readings to get acquired by the computer, and then turn the steering wheel towards the right side (+10°) and wait for reading acquisition and again move the steering wheel back to **Steering Wheel Straight Ahead** position. Using the data acquired during this operation, the computer calculates the Caster angle and Kingpin angle.

Initial Readings of Wheel Alignment

After measuring the caster and kingpin values, the steering wheel must be stopped in a straight-ahead position. At this position the camber and toe readings of all the wheels are acquired by the computer. Also the run-out compensation and THA compensation are applied properly, and all the initial alignment parameters of the vehicle are calculated and displayed on the screen, as shown in Figure 5.14. Now a **printout** can be taken, if needed.

FIGURE 5.14 Initial readings display.

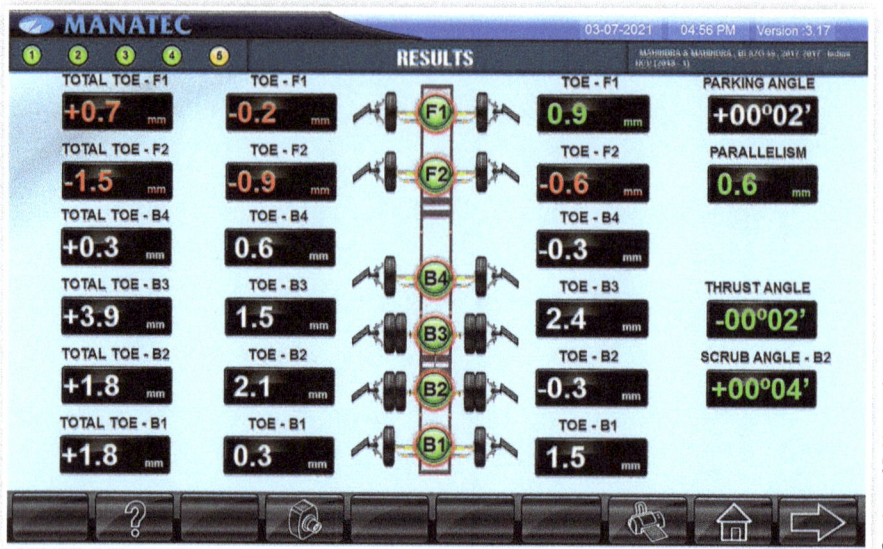

HCV Angle Corrections

In the display as seen in Figure 5.14, the angles that are out of specifications will be displayed in red color. These red-colored parameters must be adjusted and corrected in the vehicle to become green in color. While effecting corrections, the red color bar slowly moves towards the green color. When the specified value is reached, it fully becomes green in color. Now the lock nut must be tightened firmly to complete the correction. In HCVs, only the toe angle can be corrected; all other angles cannot be corrected.

Note: In certain HCVs, the camber and caster angles are correctable using special shims or eccentric cam type of shims. Some Wheel Alignment Computers offer the facility to carry out these kinds of shim corrections also.

Types of HCV Axles

Before effecting corrections, we must understand the type of mechanism available in the vehicle for effecting the corrections. In HCVs, for Wheel Alignment purposes, two types of axles are available based on the method of tilting and adjusting the wheels' angles. These two configurations are distinct from each other and are explained here. Irrespective of the suspension and axle designs, these types are universal, and it is essential to know about these two types.

Axles with Single Tie Rod or Track Rod

FIGURE 5.15 HCV axles with a single tie rod.

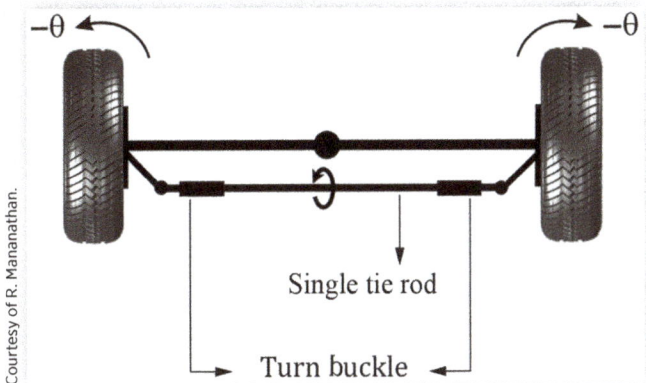

In this configuration (Figure 5.15), the left and right wheels will be connected by a single tie rod or Track Rod. The tie rod has opposite threads on the left side and the right side (Turnbuckle). When it is rotated, the left and right wheels get tilted simultaneously either towards the inside or towards the outside. This means that if the right wheel toe gets tilted by $+\theta$ degrees, the left wheel toe also gets tilted by a value of $+\theta$ degrees equally. The following axles will be having a Single Tie Rod or Track Rod for wheel angle adjustments:

- Front Axles F1, F2.
- Pusher Axle.
- Tag Axle.

Rigid Axles

FIGURE 5.16 Typical rigid axle.

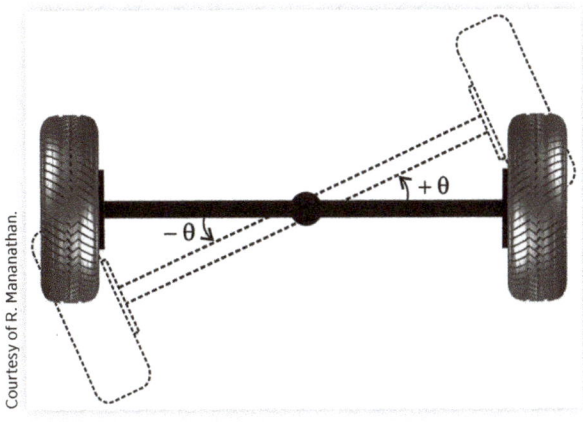

In this type the left and right wheels are rigidly mounted to the axle ends. During Wheel Alignment, if the wheels must be tilted for changing the toe angles, the entire axle must be tilted as shown in Figure 5.16. While tilting, if the right wheel gets tilted by $+\theta$ degrees, the left wheel gets tilted by $-\theta$ degrees. In this type, the **total toe** always remains the same and cannot be changed.

Correction Procedure

As explained earlier, the drive axle must be corrected first, and all other axles will be corrected with respect to the drive axle:

Step 1: Make F1 Left toe = F1 Right toe by adjusting the steering wheel.

Step 2: Apply steering lock.

Step 3: Now verify the THA. If it is not equal to zero, it must be made zero by tilting the **B3 axle.**

FIGURE 5.17 Jack position for THA correction.

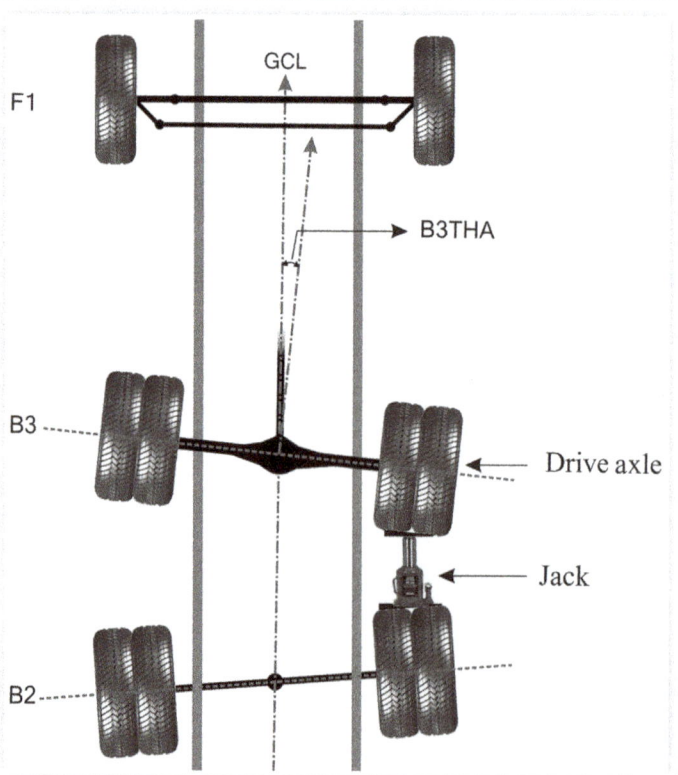

In a multi-axle vehicle, the technician must find out the drive axle which gets power directly from the engine. In **Figure 5.17**, B3 is shown as the drive axle. Since the drive axle is a rigid axle, the individual wheels cannot be tilted for correction and the whole axle has to be tilted. While tilting the axle, if the **right toe** increases, the **left toe** will decrease by the same amount and the total toe remains constant. This means the axle must be tilted till the left toe and right toe of the wheels become equal to **total toe/2**.

This situation is well represented by the THA. If the left toe and right toe are equal, then the THA will be zero; otherwise, a value will be displayed. Therefore, if the THA is made zero, then the drive axle is aligned properly.

As shown in **Figure 5.17**, a jack can be used suitably for tilting this axle and bringing the THA to zero value. While using the jack, keep the B2 axle U bolts in firmly tightened condition and loosen the B3 U bolts and apply jack pressure until the THA becomes zero or within limits.

When the THA is not exactly zero, but within limits, the color code green is displayed. Now there will be a small residual THA and a Thrust Line will exist. It must be noted that further alignment of other axles will be done with respect to this residual Thrust Line only.

Step 4: Correcting the F1 axle total toe (if the total toe is incorrect).

We made the F1LX = F1RX in Step 1. Now, loosen the lock nuts given in the single tie rod and slowly rotate and adjust the single tie rod till the correct total toe value is achieved. The single tie rod is connected to the left and right wheels by a turnbuckle mechanism, which turns the left and right wheels in equal amounts in the opposite directions, thus maintaining the equal value of the left and right toes.

Note: Before starting F1 correction, when the Left and Right toes are made equal by adjusting the steering wheel, the steering wheel is supposed to be in a straight-ahead position. But in some vehicles, the steering wheel may not be in a straight-ahead position. If the steering wheel is required to be in a straight-ahead position with equal left and right toes, then the steering wheel must be removed and refixed suitably.

Now the F1 wheels correction is over.

Step 5: Correcting F2 axle for total toe and parallelism.

This correction is required only when the F2 total toe is incorrect and the parallelism is incorrect. The F2 axle is normally connected with the F1 axle by a **Drag Link**, as shown in **Figure 5.18**. The purpose of the drag link is to ensure the F2 wheels follow the same turning radius when the F1 axle turns in a curve. This ensures that the F2 wheels are not subjected to any scrubbing action.

FIGURE 5.18 F1 and F2 axles (Twin Axle).

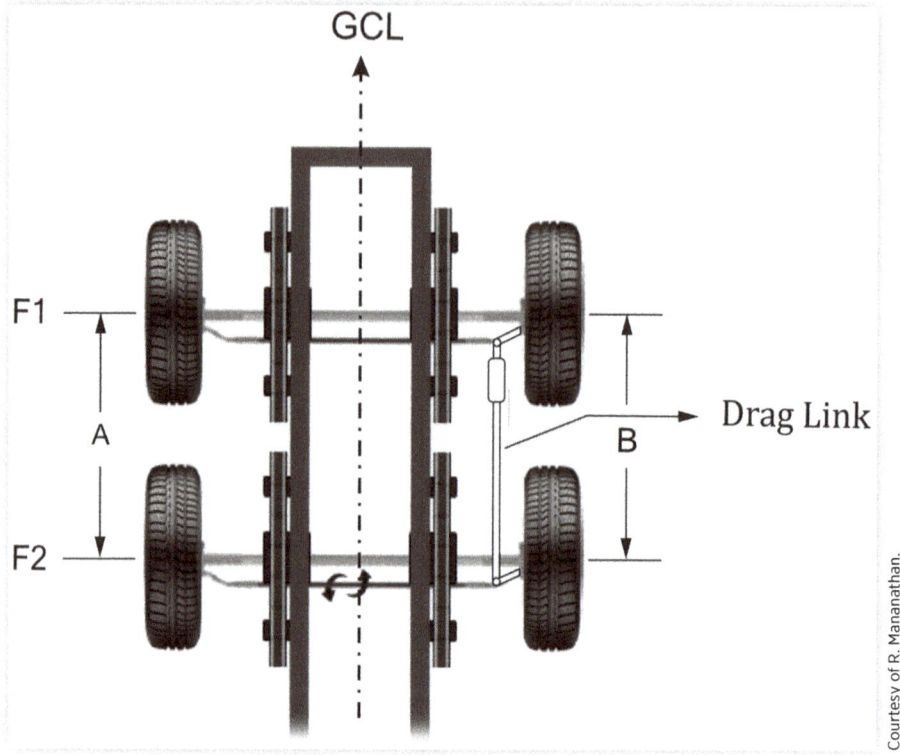

Steps to be followed for achieving the parallelism and total toe of the F2 axle:

- Turn the steering wheel and make F1LTO = F1RTO.
- Verify whether the F2 total toe is within limits. If not, using the single tie rod of the F2 axle, adjust and bring the total toe to the specification or within limits. Now, at this point, the F2LTO may not be equal to F2RTO and the parallelism may not be zero.
- To set this right, use the turnbuckle mechanism in the drag link and tilt the F2 axle till the parallelism achieves zero or within limits.

Now the F1 and F 2axles will be parallel to each other, and the left and right toe values will be equal, keeping the total toe within limits. In this condition, parallelism will be within limits.

Now the F2 wheels correction is over.

Important Note about Parallelism: Instead of achieving **Parallelism = 0**, the F1THA and F2 THA can be brought to the **within limits** condition. But in this case, there is a possibility of the F1 bisector being positive + and F2 bisector being negative – with respect to the GCL and still be within limits. Under such situation, the scrubbing action on the F2 wheels will be high and the tire wear will be abnormal. Therefore, making the F1THA and F2THA within limits, and still bringing both **in line**, is called **parallelism**. This minimizes the scrubbing between the F1 and F2 wheel to a great extent.

Step 6: Tilting and correcting the B2 axle to make the Scrub Angle zero (if the Scrub Angle is not within limits).

FIGURE 5.19 Jack position to correct the scrub angle.

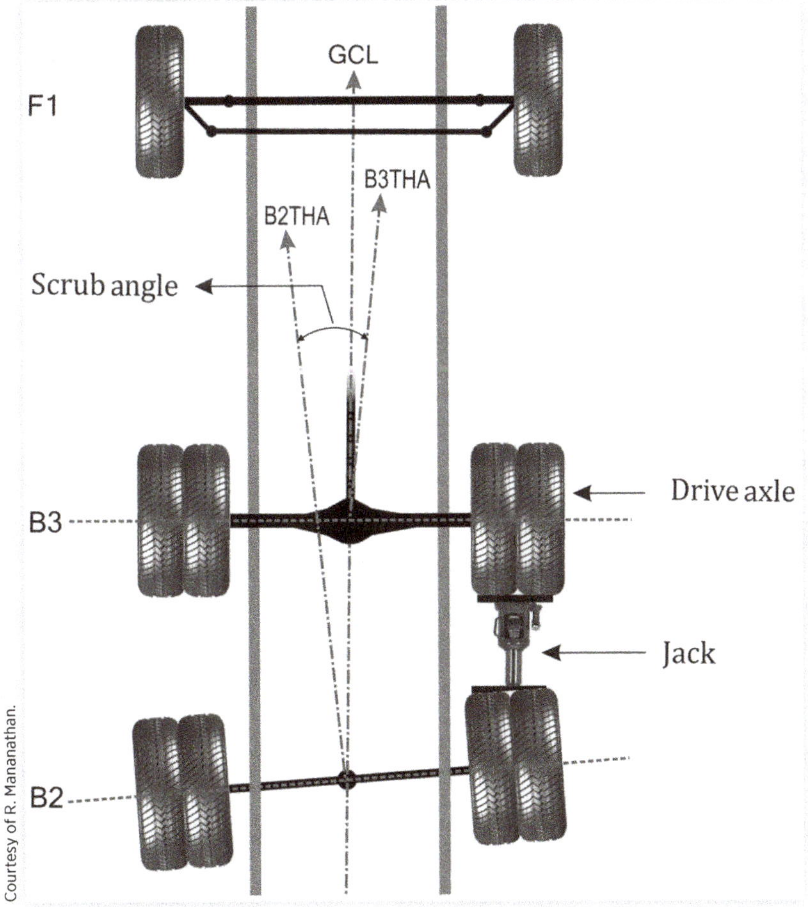

Refer **Figure 5.19**: Scrub Angle is the angle between the Thrust Line of B3 and the bisector of B2 wheel angles. The scrub angle is calculated and displayed on the screen. Now the scrub angle has to be made zero by tilting the B2 axle. This can be done using the jack and positioning it on the proper side as guided by the system.

Loosen the U bolts of the B2 axle. Using the jack suitably on the correct side, tilt the B2 axle until the scrub angle becomes zero or within limits. In practice, the scrub angle may not become an exact zero. If it reaches within limits, it is sufficient.

Step 7: Correcting Tag Axle and Pusher Axle (total toe).

While coming for alignment, the wheels of the Pusher Axle and Tag Axle may not be touching the ground. For carrying out alignment, the wheels must be lowered to make them rest on the floor and bear the load. Then the total toe readings must be observed on the screen, and if deviating from specification, it must be corrected and brought within the specification by adjusting the single tie rod.

The Tag Axle and Pusher Axle are designed to be floating axles. These are not firmly tightened to the chassis. They are designed in such a way that they travel along the Thrust Line of the drive axle. Therefore, it is enough if the total toe is set as per specification for these axles. The scrub angle will automatically become zero when the vehicle runs on the road.

Special Tools Used for Rigid Axle Corrections

In some vehicles, the wheel distances may be more and the jack may not get a supporting wheel. Under these circumstances, a **Single Axle Pusher**, as shown in Figure 5.20, will be useful.

Single Axle Pusher

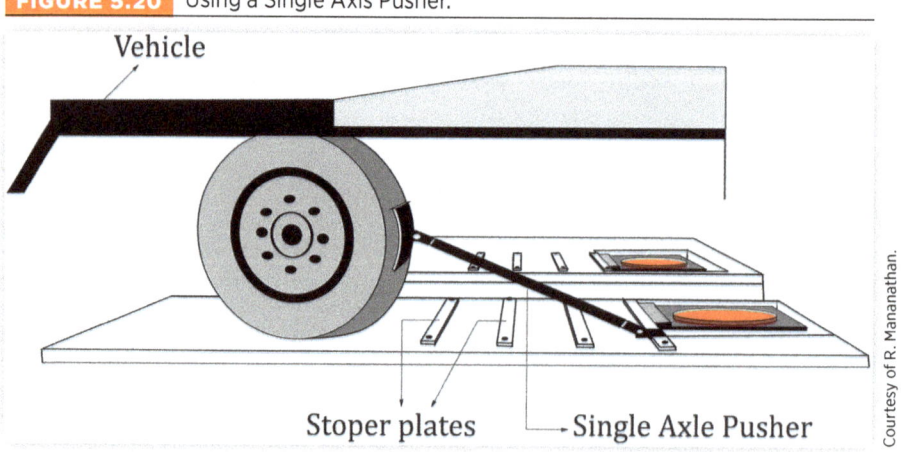

FIGURE 5.20 Using a Single Axis Pusher.

To use the **Single Axle Pusher**, stopper plates must be fixed on the floor in many places to accommodate different vehicle designs. This is a tough job. Instead, a Chain Block can be used as shown below.

Chain Block

As shown in Figure 5.21, the axle can be pulled from one side to tilt the axle and effect corrections.

FIGURE 5.21 Chain block.

Chassis or Frame Alignment

FIGURE 5.22 Chassis with six axles.

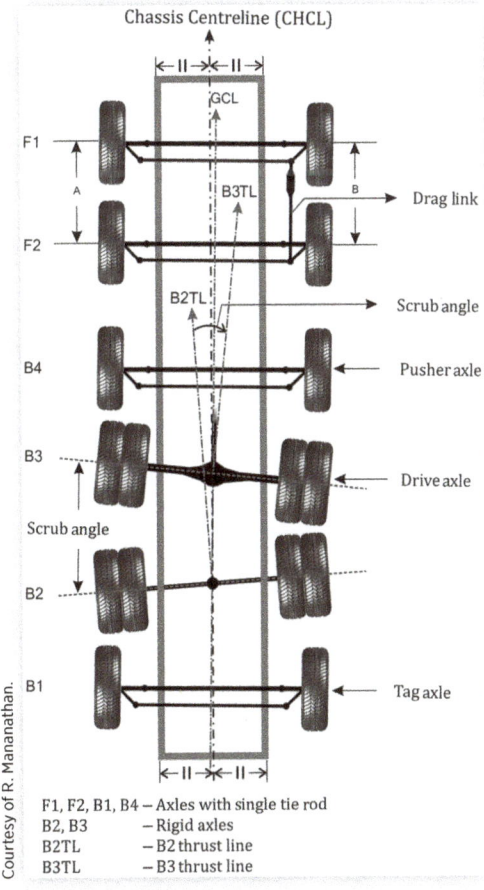

F1, F2, B1, B4 — Axles with single tie rod
B2, B3 — Rigid axles
B2TL — B2 thrust line
B3TL — B3 thrust line

As seen in Figure 5.22, the chassis of the HCVs are rectangular in shape. The line connecting the middle point of the front connecting frame and rear connecting frame of the chassis is called Chassis Centerline (CHCL). Since the HCV chassis is very long, sometimes, during fabrication, certain distortions may take place and the rectangular shape may differ a little. Still the line joining the front center point and the rear center point is called CHCL.

Normally, in a vehicle, the GCL must coincide with the CHCL. For this, the F1 axle has to be carefully mounted onto the chassis in such a way that the GCL coincides with the CHCL. (A procedure to check this is given separately.) Similarly, all other axles must be mounted parallel to the F1 axle, maintaining the F2LD = F2RD, B3LD = B3RD, etc.

If, for any reason, the F1 axle that has been mounted is a little inclined to the CHCL, then both the GCL and CHCL will separately co-exist in the vehicle geometry. In the above situation, a question arises whether the alignment must be carried out with respect to the GCL or CHCL. If alignment is carried out with respect to the GCL, the vehicle may travel in a squinted direction, as shown in Figure 5.23.

FIGURE 5.23 Vehicle travelling in a squinted direction.

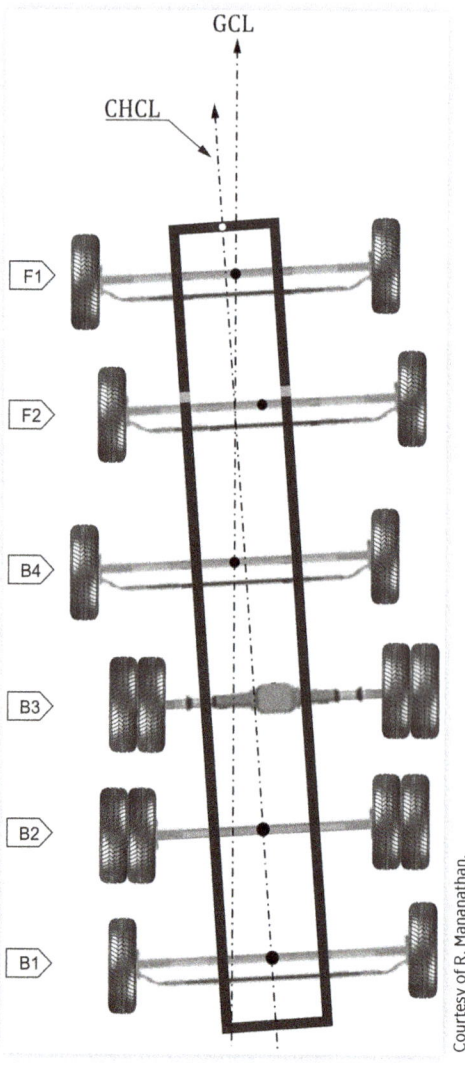

Squint travel will affect the **travel appearance**. It will not affect the Wheel Alignment and there will not be any abnormal tire wear. Instead, if alignment is done with respect to the CHCL, such squint travel appearance will not be there. But aligning a vehicle with respect to the chassis requires extra data of the chassis to be given to the system using a **Frame Reference Gauge**, and it takes extra time. Since such inclined mounting of F1 axles is rare, it is generally decided to carry out alignment with respect to the GCL. After alignment, if a vehicle travels with a squint, it can be corrected by redoing the alignment with respect to the CHCL instead of spending extra time in all the vehicles for doing chassis alignment. Nowadays, the HCV Wheel Aligners have the facility in the program to choose either GCL alignment or CHCL alignment. However, the decision of choosing the type of alignment is left to the user of the alignment computer.

F1 Axle Adjustment to Make GCL Coincide with CHCL

After alignment, if **squint travel** is noticed, it will be because the F1 axle is not mounted properly on the chassis. This can be verified as given below.

In Figure 5.24, if A is equal to B, then the CHCL and GCL will be one and the same. If not, it has to be corrected before proceeding with the alignment. Since, as an assembly procedure, it is ensured that A = B at the manufacturing stage itself, the deviation possibilities are less. Only when **squint travel** is noticed after alignment, either F1 axle proper wheel mounting can be checked and corrected or **chassis-based alignment** can be done. It is advised that such corrections be done in the workshop instead of carrying out in Wheel Alignment centers.

FIGURE 5.24 Verification of wheel mounting in the front axle.

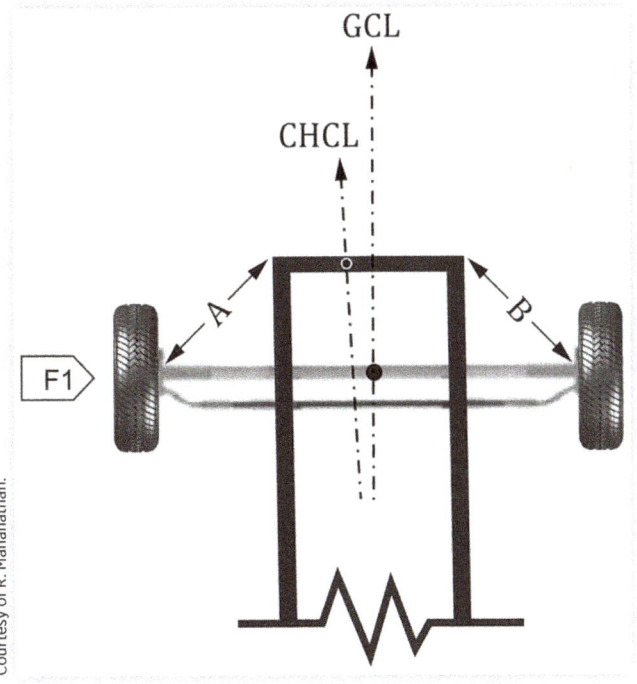

Courtesy of R. Mananathan.

Trailer Alignment

FIGURE 5.25 Jeep and trailer.

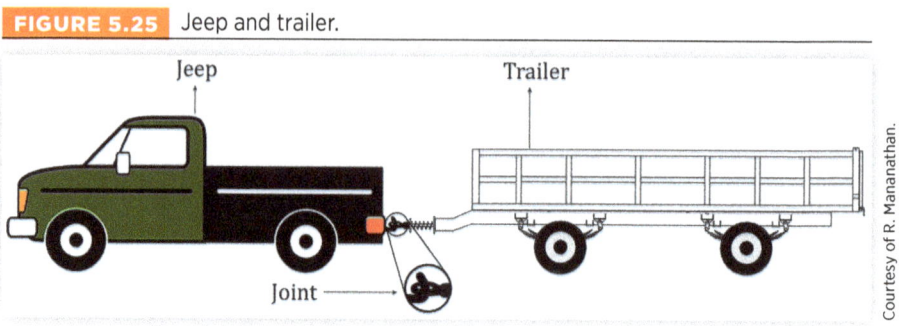

In the case of trailers, the trailer and the driving jeep are not firmly coupled with each other. They are two separate entities as shown in Figure 5.25. Both must be aligned individually. Carrying out the alignment of the Jeep is like carrying out the alignment of any other vehicle depending on the number of wheels of the jeep.

In a trailer there is no drive axle. Therefore, all the axles of a Trailer have to be aligned with respect to its front axle. All the axles in a Trailer will be rigid axles. Trailers are manufactured with three to eight axles in different wheelbase distances. Though all these configurations can be aligned by a 3D Wheel Aligner, because of the long chassis, the image plates may go far away from the visibility of the cameras. Under such circumstances, the alignment can be carried out with reverse parking. Some good Wheel Aligners also offer this reverse parking alignment facility.

6

Wheel Alignment—Points to Observe

Contents

Wheel Alignment—Precautions 107
Wheel Alignment—Troubleshooting 111
When to Carry Out Wheel Alignment 112

Wheel Alignment—Precautions

Although the Wheel Alignment machine is a computer, the following manual activities must be carried out by the technicians to get the proper Wheel Alignment:

- Parking the vehicle for Wheel Alignment.
- Positioning the Rotary Plates.
- Fixing the Wheel Clamps on the wheels.
- Fixing the Steering Wheel Lock.
- Pushing or raising wheels for the run-out program.
- Fixing the Brake Pedal Lock.
- Turning the steering wheel for measuring the Caster and Kingpin.
- Correcting the incorrect angles.
- Keeping the steering wheel **Straight Ahead** wherever required.
- Periodical maintenance of the accessories.

All these activities require proper training. Apart from good training, the following precautions will make the technicians carry out Wheel Alignment successfully in all kinds of vehicles. These precautions are applicable for LCVs and also for HCVs.

Parking the Vehicle for Alignment

The vehicle must be parked either on the pit or on the lift by slowly driving the vehicle onto the center of the pit or lift. In the case of lifts, care should be taken to place the wheel stoppers to avoid overshooting of the vehicle during driving and parking and also while pushing for run-out measurement.

Positioning the Rotary Plates

The vehicle must be parked in such a way that the front wheels rest on the middle of the Rotary Plates (Figure 6.1). If not, the vehicle has to be either reparked or lifted so that the Rotary Plates can be moved and placed properly, thereby facilitating the wheels to rest on the middle of the Rotary Plates. The Rotary Plates, regardless of whether placed on the floor or on the lift, must be placed on a level surface without undulations to avoid any tilting. The top surface of the spacers given at the periphery of Rotary Plates must be on the same surface level as the rotary plate. While pushing the vehicle by 30° or 180° for run-out operation, the rotary plate should never get tilted. Even a small tilt during a run-out operation will lead to abnormal toe and camber values.

FIGURE 6.1 Rotary plate.

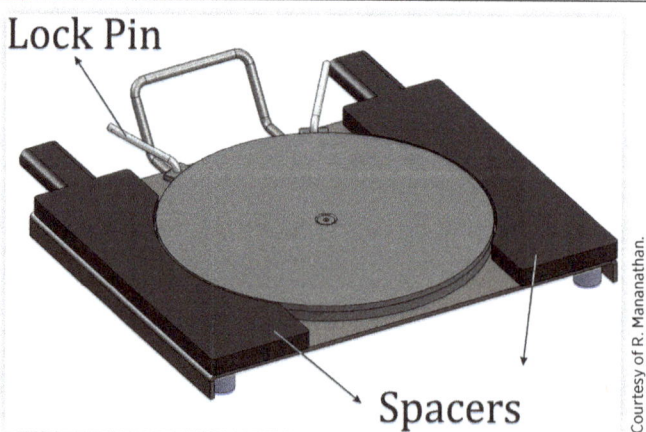

Before starting the **pushing** of the vehicle for run-out, the following precautions must be taken.

The **Lock Pin** of the rotary plate must be inserted properly to avoid any rotation.

The Steering Wheel Lock must be fitted onto the steering wheel firmly to avoid the wheels from turning during the pushing operation.

Fixing the Wheel Clamps on the Wheels

While fixing the wheel clamps onto the wheel rims, the clamps must be fixed by choosing the right spots on the rim. These spots should not have any bend or distortion. Also all the four clamps must be sitting evenly on the wheel rim without any gap. This is a manual activity to be done every time the alignment is carried out. Therefore the technicians must observe this point carefully (Figure 6.2).

FIGURE 6.2 Wheel bracket mounting.

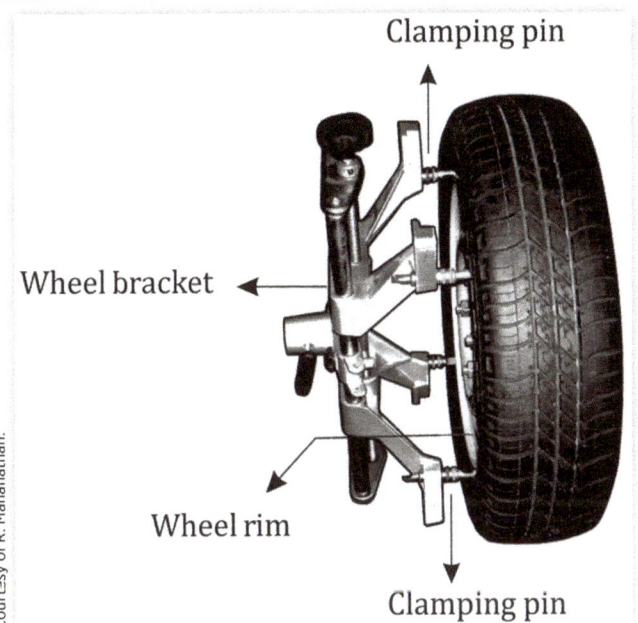

HCV Wheel Rims
Since the wheel rims of HCVs are large in diameter and are carrying huge loads, the wheel rims will be more distorted compared to the LCV wheels. In HCVs, sometimes the tires are removed using long tire levers and bars, spoiling the wheel rims. Therefore, the possibility of wheel rim distortion is high in the case of HCVs, and the technicians must be more careful in choosing the correct spot for fixing the wheel clamps.

Fixing the Steering Wheel Lock
The Steering Wheel Lock supplied along with the Wheel Alignment computer must be properly fixed to the steering wheel before the run-out measurement and in other measurements as required.

Brake Pedal Lock
Before starting Caster and Kingpin measurement, the Brake Pedal Lock must be applied firmly to avoid any movement of the vehicle.

No Jerks or Shakes
While carrying out alignment, the vehicle should not be subjected to any jerk or shake. No one should lean on the vehicle.

Steering Wheel Straight-Ahead Position

The steering wheel must be kept in a straight-ahead position during the following steps of Wheel Alignment:

- Run-out measurement.
- Initial reading measurement.
- Angle corrections.
- Final reading measurement.

This must be done manually, and the technician must keep the steering wheel in a straight-ahead condition during the above operations.

Maintaining All Accessories in Good Condition

All the accessories must be periodically cleaned, serviced, and maintained properly at least once in two weeks.

No Frequent Calibration

The calibration, once done during installation, will not change afterwards unless any physical damage happens to the aligner. If the cameras or image plates are changed for service purposes, the calibration has to be done again. Therefore, calibration should not be done frequently. If there is a problem in the output reading, it may be because of other reasons like wheel clamp fixing, rotary plate mistakes, improper accessories, etc.; only after eliminating these possibilities, the calibration can be doubted.

Surface Level

The location on which the vehicle is parked for alignment must be a **levelled surface** without undulations. Particularly for HCVs, the length being more, it may be difficult to make perfect levelling. Still, it is important to maintain the level preferably using a water level or spirit level. A good concrete flooring will be the most appropriate flooring. The Rotary Plate seating place can be of granite for easy dragging of the Rotary Plates for positioning below wheels.

Safe Tire Changing to Avoid Wheel Rim Damage

To remove the tire from the rim, do not use a crowbar or Tire Levers. Tire-changing machines are available to remove the tires without spoiling the rims, and such tire-changing machines can be used for changing the tires. There are separate tire-changing machines available for LCVs and HCVs.

Wheel Alignment—Troubleshooting

Problems that may arise after Wheel Alignment and the possible causes are listed below.

Side Pulling while Driving

- If reported in all the vehicles aligned, the reason could be improper calibration.
- If reported only in some vehicles, the reason could be an improper fixation of wheel Clamps.

Excess Toe and Camber

- Rotary plate problem (tilting during run-out).
- The balls in the rotary plate might have gotten damaged.
- Improper fixing of wheel clamps.
- Steering Lock either not fixed or improperly fixed.
- Rotary plate lock pin not inserted.

Steering Cross

- If reported in all the vehicles aligned, the reason could be improper calibration.
- If reported only in some vehicles, the reason could be an improper fixation of wheel clamps.

Improper Camber

- If reported in all vehicles aligned, the problem may be in Rotary Plates.
- If the error is more on the left side and less on the right side or vice versa (of the same value), and if the problem is noticed in all vehicles, then the reason will be the level difference on the wheel seating locations between the left side and right side.
- If the error in Camber is reported only in some vehicles, the problem could be in wheel clamp fixing.

Error in Caster

- If the same amount of Caster error is shown in all vehicles, the reason could be an improper level between the front and rear wheel seating locations.

Error in Kingpin Angle

- The reasons mentioned for the wrong Camber will be applicable for this.

Error in Caster and Kingpin

- The Brake Pedal Lock might have not been fixed properly during the **Caster Swing** operation.

When to Carry Out Wheel Alignment

- It is advised to carryout Wheel Balancing and Wheel Alignment every 5,000 kms if the vehicle is going to run on rough roads; Otherwise every 10,000 kms will be apt.
- Whenever wheels rotation is done, Wheel Alignment must be done.
- When uneven tire wear between wheels is noticed.
- If abnormal tire wear is noticed.
- Whenever new tires are fitted to the vehicles.
- If the vehicle sways or behaves in an unsteady manner while driving.
- When side pulling or steering cross is noticed.
- If the vehicle meets with an accident and/or suspension parts changed.
- During running, if the tires were found worn out substantially, instead of waiting for the tires to become bald, they must be changed immediately. Otherwise, the tire may slip during high speeds or even they may burst at any time leading to accidents.

Finally, we live in a century where continuous changes take place in the vehicle technology and design. The technicians must realize that changes will also take place in alignment technology and be prepared to adapt to the new changes.

7

Types of Tire Wears and the Causes

Contents

Tire Wear in the Middle of the Tire 113
Tire Wear in the Inner and Outer Edges 114
Tapered Wear at the Outer Edge 114
Tapered Wear at the Inner Edge 115
Feathered Wear in the Inner Edge of the Tire 115
Feathered Wear on the Outer Edge of the Tire..116
Tapered Intermittent Wear on the Inner Side of the Tire 116
Patch Type of Wear on the Tire Surface 117

We all know that the tires will wear out while running on the roads. If the wear is uniform around the periphery of the wheel, the life of the tire will get prolonged. Instead, if the tire wears more in one part of the tire than other places, then it may lead to early replacement. Depending on the location of where the excess tire wear is noticed, we can find out the reason and try to take suitable action to eliminate the cause and extend the tire life. The possible reasons for various tire wear patterns are listed below.

Tire Wear in the Middle of the Tire

FIGURE 7.1 Middle wear.

Courtesy of R. Mananathan; Tire: Courtesy of Point-S Development.

Tire pressure that exceeds the specified pressure causes tire wear in the middle of the tire (Figure 7.1).

©2022 SAE International

Tire Wear in the Inner and Outer Edges

FIGURE 7.2 Edge wear.

Tire pressure lower than the specified pressure causes tire wear in the inner and outer edges (Figure 7.2).

Tapered Wear at the Outer Edge

FIGURE 7.3 Tapered wear on the outer edge.

Tapered wear at the outer edge is a result of excess Camber when the specification is **positive** in value or less Camber when the specification is **negative** in value (Figure 7.3).

Tapered Wear at the Inner Edge

FIGURE 7.4 Tapered wear in the inner edge.

Tapered wear at the inner edge is a result of excess Camber when the specification is **negative** in value or less Camber when the specification is **positive** in value (Figure 7.4).

Feathered Wear in the Inner Edge of the Tire

FIGURE 7.5 Inner feathered wear.

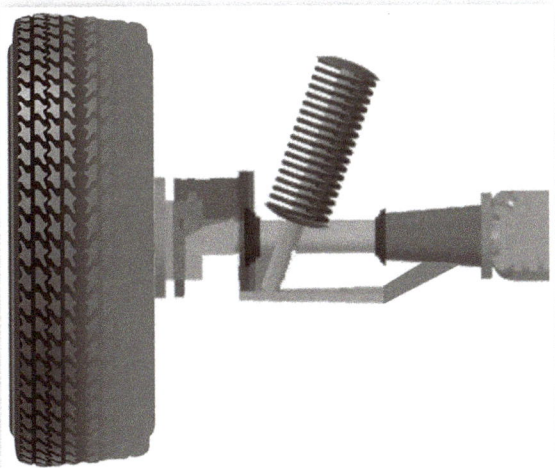

Feathered wear at the inner edge is a result of excess Toe when the specification is **positive** in value or less Toe when the specification is **negative** in value (Figure 7.5).

Feathered Wear on the Outer Edge of the Tire

FIGURE 7.6 Outer feathered wear.

Feathered wear at the outer edge is a result of excess Toe when the specification is **negative** in value or less Toe when the specification is **positive** in value (Figure 7.6).

Tapered Intermittent Wear on the Inner Side of the Tire

FIGURE 7.7 Inner intermittent wear.

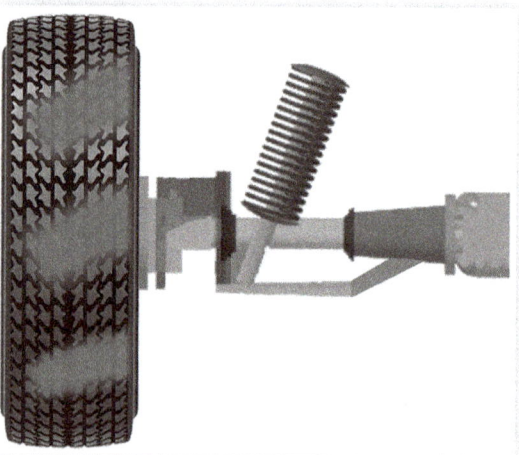

Damaged suspension parts will result in tapered intermittent wear on the inner side of the tire (Figure 7.7).

Patch Type of Wear on the Tire Surface

FIGURE 7.8 Patch wear.

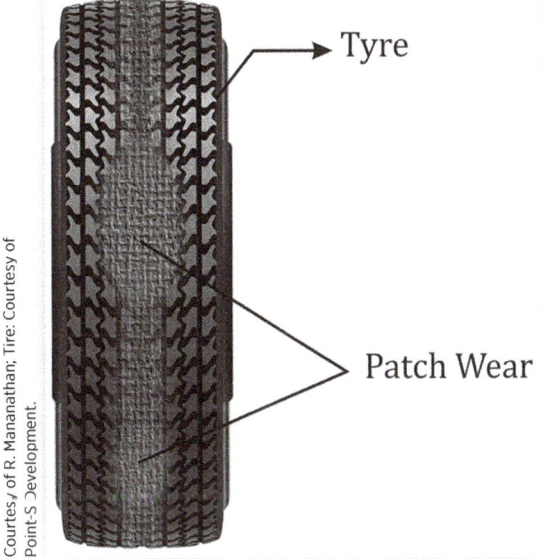

Improper wheel balancing will result in patch type of wear on the tire surface (Figure 7.8).

 Note: On many occasions, more than one type of tire wear will be seen in the tires. Still timely Wheel Alignment and Balancing will reduce further wear.

8

Tire Safety

Contents

Tire Pressure 119
Tire Rotation 121
Driver's Ability 124

Tire Pressure

Every vehicle manufacturer has specified the correct pressure to be maintained in the front and rear wheels. This is normally given in the User Manual of the vehicle (Figure 8.1). These pressures must be maintained in the vehicle at all times.

FIGURE 8.1 Tire pressure chart.

TYRE	SIZE	COLD TYRE PRESSURE
FRONT	155/70 R13 TL	33 PSI
REAR	155/70 R13 TL	34 PSI
SPARE	155/70 R13 TL	34 PSI

ALWAYS VERIFY THE RECOMMENDATIONS OF THE TYRE MANUFACTURE

Courtesy of R. Mananathan.

Normally, the front wheels will have one value and the rear wheels will have another value as specified tire pressure. In any case, it is very important to ensure that the tire pressure in all the wheels is maintained as per specified values.

The reason being both excess tire pressure and less tire pressure than the specified pressure will result in abnormal tire wear.

Effect of Excess Tire Pressure

If the tire pressure is more than the specified value, the road contact area gets reduced. This increases the road contact pressure per square inch (psi). Because of this higher psi, the tire gets squeezed and wears out fast.

Effect of Low Tire Pressure

On the contrary, if the tire pressure is less, the road contact area increases, and due to this, the friction increases, resulting in tire wear. Besides tire wear, the tire will get heated up more.

In the summer seasons, if long-distance travel is undertaken, the tire temperature may increase abnormally leading to tire burst. This is more relevant in tropical countries where the summer temperature is very high. Therefore, it is very important to check the tire pressure at least once in two weeks and maintain the correct tire pressure.

Note: If a vehicle is overloaded than its specified load, the tire pressure increases, leading to rapid tire wear. Also all the suspension parts get overloaded and are subjected to excess stress and strain. Therefore, overloading a vehicle must be avoided at any cost.

Tire Pressure Monitoring System

Considering the importance of Tire Pressure, a few developments have taken place.

FIGURE 8.2 Tire pressure monitoring system.

Courtesy of R. Mananathan.

The Tire Pressure Monitoring System (TPMS) (Figure 8.2) senses the pressure of all the four wheels at any point of time and displays it in the dashboard. This is available as a **stand-alone** unit that can be kept on the dashboard. This also comes along with the vehicle as an integrated display in the dashboard. Whenever the tire pressure in any particular wheel crosses the limit, a **beep** sound is given, indicating the wheels have excess or low pressure. Similarly, even if the temperature of the wheel increases abnormally, the TPMS warns through a beep sound. This information can be made available in the owner's mobile phone also through an App. Apart from saving from tire wear, TPMS also prevents major accidents due to abnormal heating of tires.

Nitrogen Gas (N_2)

Normally when air is filled in tires it has been observed that 1.0 psi pressure gets reduced per month in the tires, in the normal usage. Also the 1.0 psi pressure will get reduced for every 5° temperature drop, and vice versa. This is because the air filled in the tire expands or contracts based on temperature.

Nitrogen is an inert gas. This means it does not expand or contract due to temperature. Instead of air, if nitrogen is filled in tires, the pressure in the tire will not increase due to heating up. This avoids unnecessary tire wear due to pressure changes and totally avoids the possibility of tire burst due to heating up. Nowadays, N_2 filling is available in petrol stations and Wheel Alignment Centers.

Tire Rotation

All the four tires may not wear out uniformly for the following reasons:

- The number of persons travelling.
- Their individual weights and place of sitting in the vehicle.
- The ups and downs on the road.
- The driver's driving habits.
- The road camber.

The **Road Camber** of the road will be tapered on either side, as shown in Figures 8.3 and 8.4. The vehicle runs most of the time either on the left side or on the right side of the road. Because of this, the tires are subjected to different pressures, and the Left and Right tires wear out in the edges.

When the vehicle runs on the middle of the road, the left and right wheels get loaded equally and the tire wear will be uniform. But in practice, the vehicle will run either on the left side or on the right side of the road only as shown in Figure 8.4.

When the vehicle runs on one side of the road, the tires are subjected to differential pressure and wears out differently. This can be noticed when the vehicle runs a few thousand kilometers. To compensate for this, the tires can be rotated by choosing the correct rotation method given below. By this the tire life can be enhanced.

FIGURE 8.3 Vehicle running on the middle of the road (equal distribution of weight on the left and right wheels).

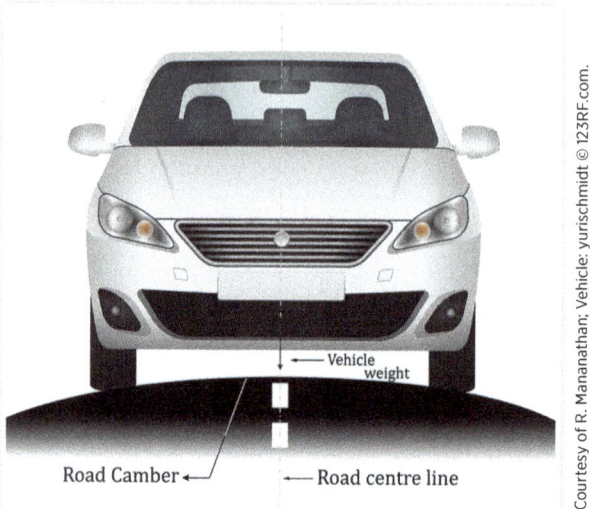

FIGURE 8.4 Vehicle running on one side of the road (unequal distribution of weight on the left and right wheels).

Rotation of Four Wheels

FIGURE 8.5 Four wheels rotation.

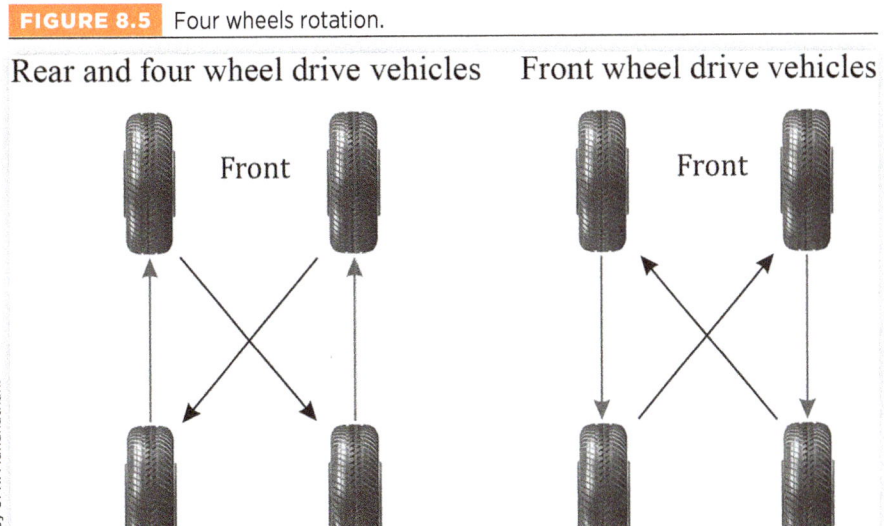

Rotation of Five Wheels Including the Spare Wheel

FIGURE 8.6 Five wheels rotation.

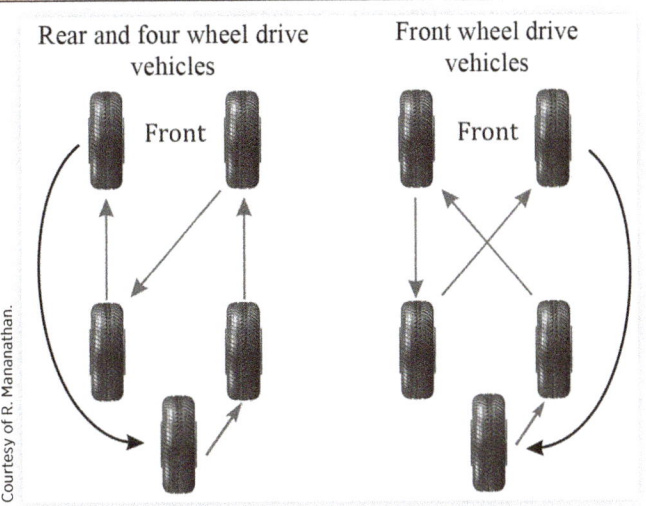

Rotation of Six Wheels

When the vehicle comes for alignment, the technician must visually inspect the wear pattern of the four wheels. If uneven wear is found, he can choose any one method of tire rotation and rotate the tires before carrying out alignment. This makes the uneven tire wear wear out uniformly, and the tire life increases.

FIGURE 8.7 Six wheels rotation.

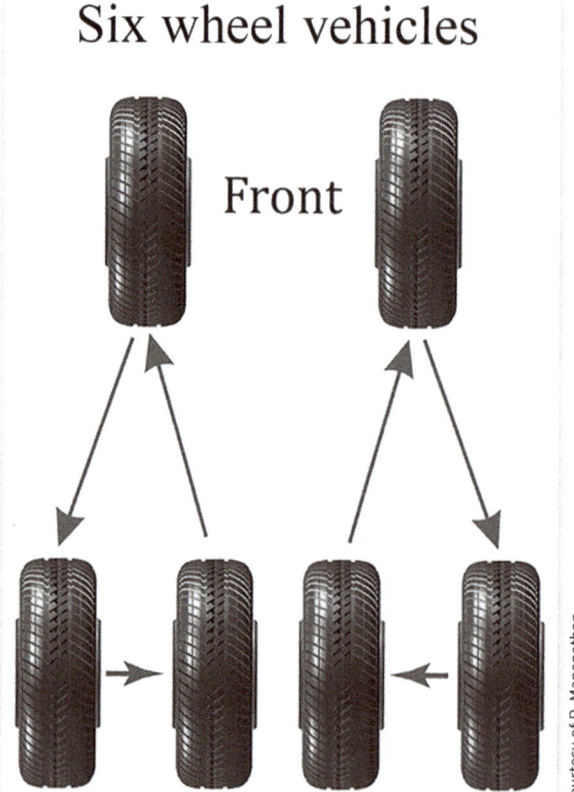

Driver's Ability

Driving habit is also a reason for excessive tire wear. The driver must avoid the following points while driving:

- Applying brake very frequently.
- Driving the vehicle in curves at high speed.
- Careless driving—unmindful of potholes or ridges on the road.
- Removing the tires using levers and crowbars.

Proper Maintenance of the Vehicle

Apart from good driving habits, the driver must also maintain the vehicle properly:

- Maintaining correct tire pressure in all wheels.
- Carrying out alignment and balancing periodically.
- Carrying out tire rotation if required before alignment.

Good driving habits and periodical maintenance can increase the tire life substantially.

9

Conclusion

The subject of Wheel Alignment and Wheel Balancing was discussed elaborately in this book. Among the two, wheel alignment is a complex subject and efforts have been made to make it simple for the readers to understand. Though the book is written for the benefit of technicians, many design aspects have been discussed on the **drive geometry** of the vehicle. This will be useful to the Design Engineers and Service Engineers as well.

Today the common man wants all the services to be done at his doorsteps. In Europe some people have started using the Wheel Balancers at home for their vehicles. In large apartments, it may be worthwhile to have this equipment installed as a common facility.

Since this book is written in simple, understandable language, every common man possessing a vehicle can read and understand the subject easily.

Knowledge sharing always gives pleasure to both the reader and also the author.

About the Author

Mr. R. Mananathan brings his vast experience in vibration and balancing technology to the readers with a straightforward style, making it easy for everyone from novices to experts to understand. Using his experience, he has designed and developed many products like Shaft Alignment computers for industrial applications, Wheel Alignment computers, and Wheel Balancing machines for automobiles. His innovative and patented development of a Wheel Aligner for multi-axle vehicles received the Innovation Award at Automechanika, Germany. Mr. R. Mananathan believes this book will enlighten the engineering community on this unique technology.

Mr. R. Mananathan is the Chairman of Manatec Electronics Pvt Ltd in Puducherry, India.

Index

A
Accessories, 14
Ackermann Principle, 32
Air compressor, 90
Aligner description, 88–89
Alignment pit, 89–90
Alignment printout, 78
Axles, 69
  adjustment, 107
  correction procedure, 98–102
  rigid axles, 97
  shift in, 87
  with single tie rod/track rod, 97
  tools, 102–103
  trailer axles, 84

B
Brake pedal lock, 55

C
Cabinet, 14
Calibration procedure, 91
Camber, 2, 83
Camber correction, 74–75
CAMs adjustment, 76
  for camber, 74
  link rod, 75, 76
  shims, 75
Caster, 3, 84, 113
  error in, 113
Caster corrections, 76–77
Caster swing, 94
Centrifugal force, 6
Charge coupled device (CCD) technology angles, 44
  sensor arm, 44–45
  wheel aligner, 44, 46
Chassis/frame alignment
  axle adjustment, 107
  frame reference gauge, 103
  with six axles, 104
  travel appearance, 103
  vehicle travelling, 105
Compensation, 41

D
Display, 14
Distortions, 37
Drive axle, 71
Drive geometry, 129
Driver's Ability, 126–127
Driveshaft, 81
Dynamic balancing, 13
Dynamic stability, 25

E
Electronic board, 14

F
Face out, 19
Factory calibration, 60
Feathered wear
  in inner edge of tire, 118
  in outer edge of tire, 118
Field calibration, 91–92
Force couple, 11
Frame reference gauge, 104

G
Geometric centerline, 24, 34, 84

H
Hairpin bends, 32
Handlebar, 30
Hard steering, 30
Heavy commercial vehicles (HCVs), 15–18
  air compressor, 90
  aligner description, 88–89
  alignment pit, 89–90
  angle corrections, 96–103
  axle configurations in, 80–81
  axle shift in, 87
  camber, 83
  caster, 84, 94–95
  chassis of, 80
  field calibration, 91–92
  geometric centerline, 84
  kingpin angles, 84, 94–95
  parallelism, 86
  parking, 92–93
  scrub angle, 84, 86
  thrust angle and thrust line, 84
  toe, 84
  and trailers, 79
  trolley jack, 90, 91
  type of axles in, 81–83
  wheel run-out, 84, 93–94

I
Inner rim, 15
Inner wheel rim, 37

K
Keypad, 14
Kingpin, 3, 113
Kingpin angle, 27–28, 113

L
Left wheel plane, 35
Left wheel toe, 84

131

Load bearing angle, 23
Lock angle, 32

M
Mass, 7
MEMS sensors, 44
Minicomputers, 44
Motor, 14

N
Negative (−) Caster, 29
Neutral plane, 40, 41
New wheels, 7

O
Old wheels, 7
Outer rim, 15
Outer wheel rim, 37

P
Parallelism, 86, 101
Parking, 92–93
Patch-type wear, 9
Patch wear, 119
Piezo electric sensor, 11
Pit, 56–58
Positive (+) Caster, 28, 29
Printout, 50

R
Residual thrust angle, 36
Resultant centrifugal force, 8
Revolution per minute (rpm), 5, 13
Right wheel plane, 35
Right wheel toe, 84
Rim, 7
Road Camber, 123
Rotary plates
  accessories maintaining, 112
  brake pedal lock, 111
  calibration, 112
  fixing wheel clamps, 110–111
  HCV wheel rims, 111
  jerks/shakes, 111
  steering wheel lock, 111
  steering wheel straight-ahead position, 112
  surface level, 112
  tire-changing, 112
Run-out, 19, 37, 38
  compensation, 50, 84
  measurement, 90

S
Scrub angle, 84, 86
Sensors, 13
Service Manual, 59
Sine curve, 40
Single axle pusher, 102
Single plane balancing, 10–11
Standstill, 1
Static balancing, 10
Steering axis angles, 22, 27
Steering wheel lock, 54, 73
Steering wheel straight ahead position, 95
Sticker weight, 16
Straight ahead position, 94

T
Tapered intermittent wear, 118
Thrust angle, 84
Thrust angle compensated, 50
Thrust line, 84
Tire pressure
  excess effect of, 122
  low tire pressure, 122
  nitrogen gas, 123
  tire pressure monitoring system, 122–123
Tire rotation
  of four wheels, 125
  Road Camber, 123
  six wheels, 126
Tire surface, 119
Tire wear
  in inner and outer edges, 116
  at inner edge, 117
  in middle of tire, 115
  at tapered wear, 117
Tire wear, causes of, 2
Toe, 3, 84
Toe angles, 24, 44
Toe correction, 73
Toe oscillation, 39
Toe parameter, 43
Total toe, 84, 98
Track width, 33
Tractional friction, 25
Trailer alignment, 107
Trailer axles, 84
Travel appearance, 104
Trolley jack, 90, 91
True vertical, 39
True vertical position, 22
Turning angle, 31
Turning radius, 31
Two plane balancing, 10, 11, 13
Two plane resolving, 12

U
Unbalanced wheels, 7–9

V
Vehicle running, 124
Vehicles and wheels
  air-filled tires, 1
  worn-out particles, 1

W
Weight addition, 11
Weight calculation, 12
Wheel alignment
  activity of, 3
  angles, 42–43
  axle with single tie rod, 70
  axle with stub axles, 70–71
  brake pedal lock, 55
  calibration, 60–62
  camber, 22–24
  camber correction, 74–75
  camera sensor, 47–49
  caster, 28–30
  caster corrections, 76–77
  CCD technology (see Charge coupled device (CCD) technology)
  color display, 72–73
  computer, 50–52
  computers, 3
  defined, 3
  drive axle, 71
  geometric centerline, 34
  image plate, 47
  included angle, 30
  installation procedure, 59
  kingpin angle, 27–28
  lift, 58–59
  lock angle, 32
  multi-axle truck for, 80
  pit, 56–58
  precautions, 109–112

problems, 3
procedure, 62–67
readings of, 95–96
ride height, 32
rigid axle, 69
rotary plate (turn table), 52–53
side pulling/steering cross, 114
steering wheel lock, 54, 73
3D planes, 46, 47
3D technology, 49
thrust angle, 35–36
toe and total toe, 24–27

toe correction, 73
toe out on turns, 30–32
troubleshooting, 112–114
wheel base and track width, 33
wheel clamp (wheel bracket), 53–54
wheel run-out, 36–40
wheel setback, 41–42
wheel stopper, 55–56
Wheel angles, 22
Wheel balancing, 129
  centrifugal force, 6
  effect of, 10

  HCVs, 17–18
  installation, 14
  problems, 2
  procedure, 14–17
  two wheelers, 18–19
  weight, 16
Wheel base, 33
Wheel clamp (Wheel Bracket), 53–54
Wheel clamping nut, 14
Wheel fixing shaft, 14
Wheel run-out, 36–37
Wheel Setback, 41–42
Wheel stopper, 55–56